Praise for The Monk in the Garden

"Robin Marantz Henig's superb book . . . tells the poignant and unforgettable story of a supremely gifted and utterly humble nineteenth-century genius, who founded an entire branch of modern science but died without ever receiving the slightest honor or recognition from his peers."
— *The Oregonian*

"[A] warm, human story of how science really works — the story, in this case, of an unassuming, mischievous, joke-telling monk whose work laid the foundation for today's race to map the human genome."
— **Erik Larson, author of *Isaac's Storm***

"Excellent . . . Henig's smart, perceptive history of Mendel and his place in scientific history is full of surprises, contradictions, strange ironies and injustices." — *Milwaukee Journal Sentinel*

"Henig's biography . . . is flavored by vivid accounts of location and atmosphere . . . She paints vivid images of [Mendel's rediscoverers] and shows why it was important for Bateson to have Mendel on his side."
— *New York Times Book Review*

"Henig has built a fascinating tale of the strange twists and ironies of scientific progress . . . Mendel's story continues to be one of the most human and appealing in the history of science, and Henig conveys its full value in this excellent and well-researched history."
— *Publishers Weekly* (starred review)

"In the tradition of *Longitude* and *The Professor and the Madman*, Robin Henig conveys the thrilling uncertainty and excitement of a great intellectual quest . . . Filled with drama, memorable characters, and genuine suspense, it's a terrific introduction to one of the most fascinating and controversial areas of modern science." — **Book-of-the-Month Club**

"What [this book] does best is to flesh out the prehistory and early history of genetics in an accessible and engaging way. It also succeeds in portraying the people behind the momentous events, warts and all." — **New Scientist**

"Compelling . . . Henig weaves together the stories of Mendel and his posthumous interpreters masterfully . . . [Her] compassion not only for Mendel but for all the other players in this drama of science shows through on every page." — *St. Louis Post-Dispatch*

The Monk in the Garden

The Lost and Found Genius of
Gregor Mendel, the Father of Genetics

Robin Marantz Henig

A Mariner Book

HOUGHTON MIFFLIN COMPANY

BOSTON · NEW YORK

First Mariner Books edition 2001

Visit our Web site: www.houghtonmifflinbooks.com.

Library of Congress Cataloging-in-Publication data
Henig, Robin Marantz.
    The monk in the garden : the lost and found genius of Gregor Mendel,
the father of genetics / Robin Marantz Henig.
        p.   cm.
    Includes index.
    ISBN 0-395-97765-7
    ISBN 0-618-12741-0 (pbk.)
    1. Mendel, Gregor, 1822–1884.   2. Geneticists — Austria —
Biography.   I. Title.
QH31.M45 H464   2000
576.5'092—dc21 [B] 00-024341

Printed in the United States of America

Book design by Robert Overholtzer

QUM 10 9 8 7 6 5 4 3 2 1

The plan of the St. Thomas monastery on page 34 is by Jessica Bryn
Henig. The diagram of a pea flower on page 70, by Maria C. Weber,
is from Alain F. Corcos and Floyd V. Monaghan, *Gregor Mendel's
Experiments on Plant Hybrids,* copyright © 1993 by Alain F. Corcos
and Floyd V. Monaghan. Reprinted by permission of Rutgers
University Press. The diagram of vinegar flies on page 244, from
The Theory of the Gene by Thomas Hunt Morgan (1926), is
reprinted by permission of Yale University Press.

In memory of my father, Sidney Marantz

*I often think of him now as one of a dying breed of men,
who want, really, nothing for themselves, who have
effaced their innermost desires without self-flagellation,
and — in order to avoid the desperations of solitude —
have given themselves over completely to their wives and
to their children, and ultimately to their children's children,
and done it with a magnificent serenity.*

— "The Flower Garden," David Guterson

CONTENTS

ACT TWO

The Monk in the Garden

Prologue: Spring 1900

There is a kind of immortality in every garden.

— *Stillmeadow Daybook,* Gladys Taber, 1899–1980

THE BLUE LOCOMOTIVE of the Great Eastern Railway streaked through the Cambridgeshire countryside. To a farmer nearby, the train's cars were a rumble of teak and steel plowing through his fields, where seedlings of barley, wheat, and oats etched their own green tracks in the springtime loam. It was May 8, 1900, and the earth, like the new century itself, pulsed with possibilities.

Among the train's passengers was William Bateson, a large, stoop-shouldered man, a don at St. John's College, Cambridge. His tweed vest strained at the buttons, his handlebar mustache gleamed; only his droopy eyes saved him from looking self-satisfied or smug. Bateson, a zoologist, had just turned forty and was one of Britain's chief combatants in the controversy over evolution and the theory of natural selection, still a source of strident debate more than forty years after Charles Darwin first proposed it.

When he boarded the train, he could have had no idea that in the next sixty minutes he would read a paper that would change the course not only of his own career but of our understanding of the place of mankind in the great cacophony of nature.

Beyond the windows of the velvet and leather compartment Bateson could see mazes of hedgerows to the left, a pretty little river to the right. A tan stucco pub, looming beyond a hillock just past Harlowtown, marked roughly the halfway point on the familiar trip from Cambridge to London. But according to the legend that has persisted for a full century, Bateson spent most of that train ride immersed in an old article from a small journal published in Austria. He was not gazing idly at the scenery.

The article, written by an obscure monk named Gregor Mendel, described the elegant botanical experiments he had conducted in a modest monastery garden in Moravia. Mendel had painstakingly crossed and backcrossed pollen and egg cells from the common pea plant to reach a better understanding of inheritance. After working on peas and other plant species for seven long years, Mendel had recorded and analyzed his findings in a two-part lecture to a local scientific society in 1865. The lecture was later published as a forty-four-page article in the society's *Proceedings* — and then was all but ignored for the rest of Mendel's life.

What had brought Bateson to that article on that morning in May of 1900 was the work of three other scientists, one of whom was the subject of his lecture that very afternoon. All three had cited Mendel's forgotten paper almost simultaneously in separate publications. Uncannily, like a field of oat stalks that somehow know to erupt in unison, the three articles had appeared within two months of one another, during that same spring.

As he read, Bateson realized that what he was trying to do in his own experiments was almost precisely what Mendel had already done thirty-five years before. He was both shocked and elated. As his wife put it, using a metaphor that prettily evoked Mendel's garden, it was as though, "with a very long line to hoe, one suddenly finds a great part of it already done by someone else. One is unexpectedly free to get on with other jobs."

By the time the Great Eastern Railway train pulled into Liverpool Street Station, Bateson knew he would have to rewrite the lecture he was about to deliver to the Royal Horticultural Society on "problems of heredity as a subject for horticultural investigation."

He had planned to focus on the work of Hugo De Vries, the great Dutch botanist whose new "mutation theory" might account for the large-scale variations that Bateson believed were necessary to propel Darwin's natural selection, the underlying mechanism of evolution. But now, as he pushed through the crowd in search of a carriage to the curving street known as Buckingham Gate, Bateson was suddenly more interested in describing the work of this unknown monk, whose findings resonated so beautifully across the thirty-five years and eight hundred miles separating London from the hilly recesses of southern Moravia.

Settling into a carriage, absentmindedly fingering his vest to be sure it was still buttoned — his wife accused him of being so indifferent to his attire that he would wear gardening clothes to town and town clothes in the garden — Bateson began to mull over his opening lines. How should he introduce this forgotten genius to the English-speaking world?

In a drafty space in Drill Hall, Bateson gave a lecture that demarcated a turning point in his evolution as a scientist. "An exact determination of the laws of heredity will probably work more change in man's outlook on the world, and in his power over nature, than any other advance in natural knowledge that can be foreseen," he began. "There is no doubt whatever that these laws can be determined."

Bateson spoke for more than an hour. Whatever his exact words that afternoon — all we have now is a text printed two years later, no doubt edited and amended to include more references to Mendel — we can surmise, based on a report published that week in the RHS's official journal, *Gardeners' Chronicle,* that there was not much discussion. But the die was cast: William Bateson had aligned himself irrevocably with the legacy of Gregor Mendel.

Within a few years, Bateson understood the sweep of Mendel's contribution. He made a pilgrimage to Brünn, the town where Mendel had lived and worked; had the monk's paper translated into English; coined the word "genetics"; and became the chief apostle of a new scientific discipline that represented the very apotheosis of the twentieth century. He became embroiled in a scientific controversy that pitted him against some of the greatest biologists of his day, including one who had been his best friend when they were both undergraduates at Cambridge. Indeed, the controversy would become so bitter and so personal that, when this former best friend died unexpectedly in 1906, some accused Bateson of killing him.

So much about gardening feels like a metaphor. Take weeding. The ubiquity of the weeds, their thorny tenacity, the hardiness of their buried roots, all seem to symbolize the pitfalls of life itself, the temptation to settle for the superficial fix even though we know that deep-seated problems will return later, or elsewhere, in other, hardier forms. It makes sense, then, to look to the garden for metaphors regarding who we are, who our ancestors were, and where we and our descendants are headed.

Part of the allure of Mendel as a hero of modern science is that we can picture him puttering in his garden, seeking answers to universal questions in his crops of peas. To some extent, Mendel's story is primarily the story of a gardener, patiently tending his plants, collecting them, counting them, working out his ratios, and calmly, clearly explaining an amazing finding — then waiting for someone to understand what he was talking about. It is the story of a gentle revolutionary who was born a generation too soon.

The myth that has grown around Gregor Mendel mirrors our contemporary understanding of scientific progress, discovery, and the nature of genius. It casts him as a tragic figure whose brilliance was unappreciated in his own lifetime. The legend is a familiar one — think of the creative geniuses who died unrewarded, from Melville to van Gogh — and it resonates reassuringly for those of us who also feel that our brilliance goes unnoticed. The story is this:

Mendel worked tirelessly in his garden for seven years, presented his findings of "certain laws of inheritance" during a two-part lecture in the winter of 1865, then passed into scientific obscurity — only to have his work "rediscovered" and resurrected by three different scientists simultaneously (one of whom was Hugo De Vries), working in three different countries, in the spring of 1900. The explanation usually given for this curious turn of events is that the world wasn't ready for Mendel's laws in 1865, and that by 1900, it was.

But, like a vine-encrusted garden path of crumbling bricks, the myth has been picked apart and slowly dismantled. Mendel was not even looking for the laws of inheritance, some scholars now say; he was just trying to find a way to breed better, more reliable flowers, fruits, and vegetables. His work did not fall into obscurity, say others; it was cited no less than twenty-two times, sometimes in prominent publications, before its raucous rediscovery thirty-five years later. And, most damaging to the traditional story, a few believe the priest was not a genius at all — just a conscientious amateur botanist with a special talent for crossbreeding who, like so many of us, stumbled into nearly every good thing that happened to him: his university training; his membership in a community of scholarly monks; his affiliation with a progressive scientific society; his position as high school teacher, for which he never earned certification; even his obscurity, which allowed him to persevere in his crossbreeding far longer than he might have with a reputation to uphold. The Mendel legend, say these revisionists, was created by biologists in 1900, when they were already locked in a heated debate about the mechanism and pacing of evolution. Those who, like Bateson, needed Mendel's laws to support their position turned a modest, meticulous, clever, and rather lucky monk into a scientific giant.

The truth, as is so often the case, lies somewhere between these extremes.

Our view of Mendel seems to have come full circle, from the original revisionism of thirty or forty years ago to the more respectful attitudes of today. The question now is not so much whether the man was a genius but where exactly his genius lay. He seems to have been a plodding, hard-working, single-minded sort, a genius for whom discovery was, as Thomas Alva Edison put it, "one percent inspiration and ninety-nine percent perspiration," not a playful, intuitive genius like Picasso. The great painter once said, "I do not seek — I find," an attitude that describes many of the men and women we now think of as geniuses.

In the conventional use of the word, genius is something one is born with, something that sets a person apart from ordinary humans, with our typically linear, categorical way of looking at the world. Mendel's genius was not this flamboyant, touched-by-an-angel kind. He toiled, almost obsessively, at what he did. But still he had that extra one percent, that inspiration that helped him see his results from a slightly different angle. It was this flash of insight that allowed Mendel to perform a feat of genius: to propose laws of inheritance that ultimately became the underpinning of the science of genetics. Even if he was subsequently lionized by men with their own agendas, even if he was not in fact the heroic father of genetics he was once made out to be, that should not diminish what Gregor Mendel was: a man with a vision and the dedication to carry it to its brilliant, radical conclusion.

To tell the story of Mendel's life and intellectual flowering requires some educated deduction. We have so little specific information about him — barely more than three short published papers, seven letters to a botanist in Munich, and a brief autobiography written when he was twenty-eight years old. Almost nothing exists that places Mendel in his garden, his monastery flat, his church, or his beloved orangery on any particular day.

This is, in certain ways, a good thing. "We are lucky to have limited information and are completely free," said one scientist. "We can speculate to our hearts' content because nobody can say we are wrong. They can only say, 'I do not agree with you.'" What that freedom means for this book is that occasionally I must indulge in speculation — not quite to my "heart's content," because this is not a novel, but more than most nonfiction writers are accustomed to. Though I have no way of knowing with certainty what our protagonist was doing or thinking at any particular time, I can tell his story, based on circumstantial evidence and the sifting through of scenarios, the way it most probably occurred.

Most of the myths about Mendel grew directly from the bitter fight between Bateson and the Mendelians on the one hand and Bateson's erstwhile best friend and the so-called biometricians on the other. Both sides were playing for the highest stakes: the right to claim a truthful insight into the workings of the natural world. What they uncovered eventually became the foundation of a science that has taken us to the very brink of human possibilities.

Mendel observed that traits are inherited separately and that characteristics that seem to be lost in one generation may crop up again a generation or two later, never having been lost at all. He gave us a theoretical underpinning for this observation, too: he believed the traits passed from parent to offspring as discrete, individual units in a consistent, predictable, and mathematically precise manner.

In Mendel's wake followed a steady string of discoveries: that these hereditable units can be found in the genes, which in turn are found on the chromosomes, which are in turn found in the cell nucleus. By the 1940s scientists knew that the meaningful information of the genes was packed into a molecule called DNA; by the 1950s they could build a physical model of the DNA molecule and interpret the code through which DNA talks to the cell. Since then the science of genetics has been consumed with using that code to see where genetic disorders originate and to map out, in order, all the pieces of DNA that fit together to make up an entire organism: first a bacterium, then a worm, and, in quick succession, a fruit fly, a dog, a rat, a plant (one of the simplest, *Arabidopsis thaliana*), and ultimately a human.

As this knowledge, especially about the human genome, began to unfold, geneticists started to tinker with our natural inheritance. Some of the tinkering has been controversial: the eugenics movement, which advocated selective breeding to improve the gene pool by prohibiting "misfits" from marrying; the possibility of human cloning, which would enable people — most likely the richest ones first — to produce younger identical twins of themselves; and research into genes that might carry traits a society values and for which it would pay dearly to pass on to the next generation, even superficial ones such as tallness, thinness, or protection against going bald.

The same impulses that sparked eugenics, cloning, and the search for the ideal child have led to some of the worst atrocities of the twentieth century. Barely thirty years after the word "genetics" was coined, Adolf Hitler was coalescing the Nazi party and masterminding the massive genocide program that would be his "final solution." Hitler's Holocaust cast its long, ferocious shadows into the end of the century as well, when "ethnic cleansing" campaigns were conducted in Rwanda and the breakaway regions of the former Yugoslavia. We have never gotten far from the belief that our genes are our destiny.

At the same time, the tinkering made possible by the century's explosion of scientific knowledge has been nothing short of miraculous. How astounding that we can now screen for defective genes through DNA analysis, help build families through genetic counseling, even exchange good genes for bad. And how awesome — as well as morally complex — that we may soon, if we want, be able to manipulate the fate of future generations, so that every newborn is guaranteed to be, at least as geneticists define it, perfect.

Nearly a century after the debate over Mendelism that set the stage for contemporary genetics, almost every part of our modern understanding of how the world works — the relationship between parents and offspring, the delicate interplay between identity and individuality, the underpinnings of nature and the commonalities among all living things — can in large measure be traced back to that startling spring of 1900, when anything was possible.

Act One

1

In the Glasshouse

How I love the mixture of the beautiful and the
squalid in gardening. It makes it so lifelike.

— *The Letters of Evelyn Underhill* (1875–1941)

GREGOR MENDEL was in the glasshouse again. It was the only place where he could get warm. Even on a sunny summer day the St. Thomas monastery, where Mendel lived, was always cold. It had been built in 1322 like a fortress, but its original purpose was to protect a community of Cistercian nuns. (The Cistercians, an order that dated back to the eleventh century, were a subset of Benedictines who wore white instead of black and followed stricter rules of conduct and devotion.) Long brick walls enclosed the vast property, and rolling hills and fruit trees gave it the drowsy feeling of a country estate.

Nearly five hundred years later, a different order of Catholic clerics, Augustinian monks, took over the building. Until the end of the eighteenth century, these monks had been living in lavish quarters in an ornate confection of a building in the heart of Brünn — in those days the capital of Moravia, in the middle of the Austro-Hungarian empire. But in 1793 Emperor Josef II, nicknamed "the good emperor," evicted the monks so he could use their beautiful building for his own residence and government offices. So the Augustinians moved just beyond the city limits to the nunnery, lying in disrepair, which they converted into a dwelling where they would feel more at home. They tore down the walls separating the nuns' cells so that a space that once had housed two or three sisters became a flat for a single priest.

Despite the improvements, nothing could rid the monastery of its chill. The corridors might hum with priests and lay brothers engaged in thought, study, and earnest conversation, but no amount of intellectual heat could warm the building's thick brick walls and hard floors of stone. Its surfaces stolidly kept out every ray of sunshine, retaining year round a wintry chill.

The glasshouse was different. Mendel frequently took sanctuary in the little two-room building nestled into a corner of the monastery courtyard right up against the brewery next door. It gave him not only blessed warmth but also the space to engage in his scientific pursuits — which would, he believed, prove important enough in time to earn him a place in the annals of horticulture. He had filled the glasshouse's long tables with pots of pea plants, each carefully labeled as to seed source and variety. His immediate goal was to breed these peas, thirty-four different seed types in all, after allowing them to self-fertilize for two full years. In the speeded-up growing seasons of the glasshouse, two years of growing meant perhaps six full generations — enough to assure Mendel that the seeds were indeed what they appeared to be.

He did not know exactly what species he was growing. They were all from the genus *Pisum* — the common garden pea — and he supposed that most were *Pisum sativum.* But among his thirty-four types were doubtless some examples of a few other species, among them *P. quadratum, P. saccharatum,* and *P. umbellatum.* The exact classification did not really matter to Mendel. As he said, it was "just as impossible to draw a sharp line between the hybrids of species and varieties as between species and varieties themselves." Fortunately, he considered this sharp line to be "quite immaterial" to his ultimate experimental goals.

What was important to Mendel — indeed, crucial, and a central factor in his subsequent experiments — was whether his seed stocks, whatever species they were, could "breed true." In other words, he needed to be sure that green peas would always have green offspring, and yellow peas always yellow; that tall plants would consistently give rise only to talls, and dwarf to dwarfs alone. Once he was certain that he had true-breeding strains, Mendel would be ready to begin his work. He planned to move beyond the glasshouse and to start planting peas outdoors — by the dozens, by the scores, and then, if he could find the space, by the hundreds and thousands. He intended to cross-fertilize the peas, and to make close observations, trait by trait, of the hybrids he created, following them and counting their offspring for as many generations as he could.

Mendel had not always worked with peas; at first, he had tried breeding mice. But toying with the reproduction of mammals, according to the local bishop, was simply too vulgar an undertaking for a priest. The bishop, Anton Ernst Schaffgotsch, had irritated the monks of St. Thomas for decades. The monks wanted to pursue their interests — natural science, physics, musical composition — unfettered by the restrictions of the Catholic Church. But Bishop Schaffgotsch would not allow such irreverence. He was especially bothered by the abbot, an independent thinker and a powerful man who seemed intent on running his monastery more like a university than a cloister. In June 1854, the year that Mendel began growing peas, the bishop visited St. Thomas, hoping to get a tighter grip on it once and for all. His ultimate goal was to shut down the monastery altogether.

The abbot proved too wily an adversary for Schaffgotsch, who was not an especially clever man. But the two clerics did eventually reach a compromise: the monastery could remain open as long as the abbot changed some of the things that Schaffgotsch found most offensive. Among them were the mice that Mendel kept in cages in his two-room flat, where they gave off a distinctive stench of cedar chips, fur, and rodent droppings. He was trying to breed wild-type mice with albinos to see what color coats the hybrids would have. Schaffgotsch seemed to find it inappropriate, and perhaps unnecessarily titillating, for a priest who had taken vows of chastity and celibacy to be encouraging — and watching — rodent sex.

"I turned from animal breeding to plant breeding," Mendel later said with a chuckle. "You see, the bishop did not understand that plants also have sex."

With his enforced new focus on plants, Mendel took as models the famous hybridists he had read about in his formal and informal studies of botany. His goal was not only to emulate them but to go beyond them, building on his knowledge of mathematics and on the scientific methods he had recently learned in his study of physics and chemistry. He wanted to apply the standards of these "hard" sciences to biology, at the time considered one of the "softer" sciences. Ultimately, Mendel wanted to find the laws that governed the creation of hybrids and to learn, perhaps, how hybrid plants spread their individual characteristics over the generations.

In hindsight, we cannot say exactly how grand his aspirations were. Knowing as we do now that Mendel would achieve fame as the father of genetics, it is appealing to assume that he started out with the loftiest of goals. It is appealing, too, to see him as a prescient genius who chose peas — which would prove ideally suited to his purpose — as the key to unlocking the secrets of heredity.

We will never know for certain what Mendel set out to do — or whether he completely understood what he found. He may have turned to peas with exactly the same question he had been hoping to answer with mice: how hybrids happen and what general laws about inheritance are revealed by their patterns of descent. As Mendel himself explained it, his goal was "to follow up the developments of the hybrids in their progeny." He hoped to provide an explanation for observations he and many others had made: that hybrids usually produce plants that look just like them, but occasionally produce plants that more closely resemble those of an earlier generation.

Part of what Mendel wanted, however, was nothing short of eternal fame. This, at least, was his dream as an adolescent, when he wrote a poem that showed he had his eye on posterity. The poem, written when Mendel was attending the Gymnasium in Troppau (now Opara), some twenty miles from his home, is ostensibly a paean to Johann Gutenberg, who invented movable type in the 1430s. But it also gives voice to the young Mendel's own ambitions. He was, after all, already something of a legend in the tiny hamlet of his birth, having been sent to the Gymnasium — a secondary school for students on their way to university — because he was so clearly one of the most promising students ever to have passed through the parish elementary school. Who can blame him for identifying with the brilliant inventor to whom society had finally granted some overdue acclaim?

In his Gutenberg poem, the teenage Mendel eerily foreshadowed his own fate, one he could not possibly have predicted as a boy: obscurity during his own lifetime but, after his death, a new voice that persisted and grew far into the next century and the next.

Like Gutenberg, the Mendel of posterity would, if he could peer down from some heavenly perch, bear witness to what he described in his adolescence as "the highest goal of earthly ecstasy": a permanent position as an intellectual hero.

> Yes, his laurels shall never fade,
> Though time shall suck down by its vortex
> Whole generations into the abyss,
> Though naught but moss-grown fragments
> Shall remain of the epoch
> In which the genius appeared. . . .
> May the might of destiny grant me
> The supreme ecstasy of earthly joy,

The highest goal of earthly ecstasy,
That of seeing, when I arise from the tomb,
My art thriving peacefully
Among those who are to come after me.

In the real world of Mendel's youth, though, the road to eternal recognition was often tortuous. When it was brightened by detours or victories, these usually were not because of something he had set out to do, but almost in spite of it. Of course, many lives unfold in this way, but we tend to believe that great men, genius-caliber thinkers, start out with specific goals and clear road maps to achievement. Not Mendel. He went to elementary school and from there to Gymnasium because his teachers told him to. He continued on to the local philosophical institute because that was what bright boys did, especially those unsuited for farming. He ended up in a monastery that encouraged his best ideas because one of his professors had directed him there. And he engaged in experimentation because that was what his abbot wanted him to do.

Which one of Mendel's boyhood "episodes" was the last straw? Which one convinced his mother and sisters, who were expected to cater to him whenever he took to his bed, that a pattern was developing, that maybe young Johann (as Gregor Mendel had been christened) had a problem deep inside? Was it the year he came home during summer term at the age of seventeen and stayed in bed for four months?

He is a grave disappointment to me, Anton Mendel would have had cause to grumble during those dark days of the summer of 1839. The Mendel farm was near Heizendorf, a tiny hamlet in Silesia, a German-speaking region north of Moravia carved out of what had once been Sudetenland. And the year had been an especially hard one for the Mendels. That winter Anton had been badly crippled by a falling tree. By summer he could still barely move. But after a short convalescence — all that he could allow himself — he was again dragging his aching body to the fields every dawn.

And what did his young, well-fed, able-bodied son do? He could barely get out of bed.

Anton was a small, thin, swarthy man with dark, hooded eyes and a gloomy outlook. His eldest child, Veronika, was much like him in these ways. But for all Anton's pessimism, he did what needed doing. He saved enough to buy the farm from its former feudal owner, worked the land no matter how debilitated his body became, and allowed himself — rarely, only rarely — the pure pleasure of working in his orchard and experimenting with grafts to create the juiciest and prettiest fruits.

His wife, Rosine, was the opposite. Sweet-natured, broad-browed, tending toward stoutness, she was easy to bring to laughter and always looking for the good in people. The couple's youngest child, Theresia, was much like her mother. And their middle one, the boy, was a combination of Anton and Rosine, a hybrid who inherited his mother's looks and sweet temperament and his father's gloomy pessimism.

What can be the matter with him? the elder Mendel might have hissed to his wife in May or June or July, as Johann's mysterious illness dragged on and on. All that book learning he was acquiring — at such hardship to the family — seemed only to pull Johann further and further away, almost as though he believed himself to be too good for farming. But if Johann felt distant from his family during that time, it was only because the work did not suit him — and because he was aiming toward dreams that had nothing to do with the farm. Ever since he had been sent away from Heizendorf at the age of eleven, he had felt cut out for a different life from the one his father and neighbors lived. Now that he was at the Gymnasium in Troppau, he knew with even more certainty that a farmer's life was not for him.

And didn't he suffer, too? His parents no longer paid any part of his school fees, leaving him to provide for himself entirely. How many boys his age had to piece together a scanty livelihood through private tutoring, all the while trying to keep up with their studies? How much longer could he bear the strain of it all? Far easier than to think through a solution was simply to hide out, at least for a while, in his bed.

Johann spoke not a word of his struggle to his mother, who came to his room every morning with breakfast and comforting words. He said nothing to Theresia, who was only ten years old. He barely saw Veronika, who was already married and out of the house; even had she been at home, his hard and humorless older sister was the last person on earth in whom Johann would have confided.

He did not speak about his feelings to his father, either. How could he tell Anton how much it pained him to watch the older man hobble off to the fields every morning — and yet how helpless he felt to do anything but watch? The question of why Johann could not get out of bed was in the back of everyone's mind — and in the forefront of Johann's.

Somehow the boy recovered and was able to return to school for the fall term. He graduated from Gymnasium the following spring and from there moved to the Philosophical Institute, a two-year program required of Gymnasium graduates before they could begin university studies. The institute was in the city of Olmütz, whose name in Czech, the primary language spoken there, was Olomouc. Life was even more difficult in Olomouc than it had been in Troppau. With no money, Johann was hungry and cold; his Czech was stilted, and, in large part because of the language barrier, he could not find enough tutoring assignments to keep himself afloat. He spent his time, as he described it, filled with "sorrow over these disappointed hopes and the anxious, sad outlook which the future offered him."

Once more he went home and took to his bed — this time for an entire year. By now it was 1841. The nineteen-year-old boy watched with growing self-reproach as his father struggled to his feet each day. He thought about how he wanted to live his life and whether he would ever accomplish his dreams in the face of what seemed to be insurmountable brick walls and unavoidable dead ends.

Veronika's husband, Alois Sturm, finally agreed to take over the farm, which relieved Johann of the guilt that came from failing to do so himself. But his real rescue came from his beloved younger sister, Theresia. Barely out of childhood, Theresia offered Johann a portion of her share of the family estate — the portion that was to have been her dowry — to see him through the two-year program at Olomouc. Out of lifelong gratitude, Mendel later took upon himself the financial and emotional support of Theresia's three sons, two of whom went on to become physicians. Helping to raise his nephews eventually became his most sacred obligation and one of his greatest pleasures.

Within two years of Theresia's loan, however, it became clear that no amount of generosity from his sister, even supplemented by a small scholarship and tutoring, could ensure Mendel the schooling he yearned for. "His sorrowful youth," he wrote, referring to himself in the third person, "taught him early the serious aspects of life, and taught him also to work." That is when he chose the only path available in nineteenth-century Mitteleuropa for a penniless young man in search of an education. At the urging of his physics professor at the Philosophical Institute, Friedrich Franz, who was also a priest, Mendel signed on with the monks.

These were not just any monks. Mendel's professor had sent him to a remarkable monastery, led by a remarkable abbot, in the remarkable city of Brünn. Franz himself had spent twenty years living in the St. Thomas monastery, which by 1843, when Mendel arrived, was under the abbacy of Cyrill Napp.

Franz knew that Abbot Napp — who had been his good friend during his own years there — was building a community of scholars within the monastery's whitewashed walls. And he knew Mendel would find a perfect niche in this community. He wrote to his old friend, assuring Napp that Mendel was "a young man of very solid character" who was "almost the best" among his physics students. This was a helpful gesture on Franz's part, one that would shape Mendel's future, but the words he chose did not suggest a nascent genius. Even though Mendel was the only student Franz recommended to Napp that year, calling him "almost the best" was faint praise. Was Mendel really not that extraordinary — simply a bright, competent, persistent young man who deserved a chance? Or was Franz, like so many of his contemporaries, blind to Mendel's one-track, simmering genius that had a chance to explode only years later, when the twin stars of intuition and accident were momentarily aligned in Mendel's favor, providing him an insight into the mystery of inheritance that few but he were prepared to understand?

Brünn, the provincial capital of Moravia and one of the fastest-growing cities in Europe, had at the time a population of 70,000. Like other cities in Austria, it was hobbled by a language war: the Czech majority wanted to maintain their language and culture, but the Germans in power prohibited the use of Czech in commerce and in many schools. Brünn had fine schools, a few good choral groups and orchestras, a Philosophical Institute, a new Technical University, and an impressive number of scientific societies for a city of its size. When Mendel arrived at the monastery, one of the most prominent of these was the Royal and Imperial Moravian and Silesian Society for the Improvement of Agriculture, Natural Science and Knowledge of the Country, known informally as the Agricultural Society. It had been founded by a group of amateur naturalists in 1806, and since 1827 its incumbent president had been Abbot Napp.

The St. Thomas monastery was located in the city's oldest quarter, known as Altbrünn, near the banks of the River Schwarza at the foot of a hill crested by the Spielberg Castle. The Spielberg, a glowering thirteenth-century fortress, was at the time still functioning as a prison for the most vicious criminals and the most outspoken revolutionaries. The lower levels (the "dark chambers") housed murderers, rapists, and thieves; at the higher levels were political prisoners who threatened the rule of the Hapsburg dynasty. Abbot Napp and several of his monks made regular trips to the Spielberg to look after the souls of the pitiable men imprisoned there.

The monastery itself felt more like a college dormitory than a house of God. St. Thomas was run according to the Augustinian credo *per scientiam ad sapientiam:* from knowledge to wisdom. Augustinians were among the most liberal of the religious orders within the Catholic Church, less Spartan than Benedictines, less isolated than Carthusians, more civic-minded than Premonstratensians. Benedictines, for instance, ate only one or two meals a day except on feast days and followed strict rules of discipline right down to small details of behavior: sleep in your shirt, drawers, and gaiters; take your shoes off under the bedclothes; no singing in bed. Carthusians lived in isolated cells, each with a small walled garden, bedroom, and study, and never saw or spoke to anyone, not even the lay brothers who delivered their meals, except for a few hours on Sundays. And Premonstratensians (called "white canons" because of the color of their robes, just as Augustinians were "black canons") spent most of their time working within the monastery walls.

Augustinians, in contrast, emphasized teaching and research over prayer. This was especially encouraged after 1807, when Emperor Franz I decreed that the monks of St. Thomas (joined by members of the local Benedictine and Premonstratensian monasteries) were to take over the teaching of mathematics and religion at Brünn's Philosophical Institute, which was to open the following year. There was a growing understanding in Austria at the beginning of the nineteenth century of the importance of science — an understanding that the Moravian Catholic Church shared with the secular intellectuals and aristocrats. Cyrill Napp, before he was elected abbot, was among the first to take up this calling.

What a gift this monastery was for a man like Gregor Johann Mendel. (The name Gregor had been assigned to Mendel when he joined the order, and thereafter he used his religious name instead of, or sometimes before, his christened name.) Not only was he free to learn whatever he could in all the sciences that fascinated him — meteorology, botany, physics, mathematics — but he was encouraged to, by the empire, by the Church, and by Abbot Napp. Napp's encouragement was the most direct and the most meaningful. The first of his many gifts was to grant Mendel free access to the glasshouse, the sanctuary-cum-laboratory in which we first find him in the summer of 1854. The second was to plan the construction of something even grander: a greenhouse of roughly twice the size, for Mendel's use alone. (Although "glasshouse" and "greenhouse" are usually used interchangeably, in this case the monastery glasshouse was heated by a stove, the greenhouse by trapped sunshine alone.)

Napp did not usually indulge his favorites. But he made an exception for Mendel, of whom he was especially fond. Perhaps Napp, who was almost exactly forty years older than Mendel — sixty-one in 1843 — saw in him something of himself as a younger man: eager, driven, and scientifically voracious.

The abbot was a solid citizen of Brünn. Born into a wealthy family, he held positions that often accompanied the abbacy with local banks and the provincial parliament. He also played an important role in the intellectual life of the town as director of high school education for Moravia; president of the Brünn Pomological Association (whose members raised fruit trees); and president of the Agricultural Society. He reflected the high regard in which science was held in those days, even among high-ranking figures of the Church. In his study of scientific sheep breeding, he had a clear, scientific view of the questions involved in discovering how species originate, persist, and change: "What is inherited, and how?"

Like many intellectuals in Moravia at the time, Napp encouraged scientific study and experimentation among his fellow amateur naturalists — and, most especially, among his monks. He encouraged Mendel's interest in meteorology — a subject on which the younger man lectured frequently and wrote up occasional reports — as well as in botany.

But for all the breadth of his intellect, Napp had a stodgy, imperious side. Rumor had it that he made his own mother address him as "Your Grace." He kept an emotional distance from the brethren, communicating primarily through Prior Baptist Vorthey, who was in charge of the novices. A small, slight man who walked with a limp, Napp could appear formidable when he chose to, as when he waited at the porter's lodge to intimidate a monk who had once too often returned home drunk in the wee hours of the morning. The monk, Aurelius Thaler — himself a botanist of some note and at the time the keeper of the monastery's small experimental garden — stumbled back home that night and, as he usually did, rang at the porter's lodge to be let in. There he was confronted not by the friendly porter but by the censorious abbot himself, in all his clerical regalia, glaring.

Monastery life was a balm to Mendel. Its regularity provided ease and comfort to a man who had spent his first twenty-one years in a thicket of uncertainty. Twice a day, in the early morning and again in the evening, services were held at St. Mary's Basilica, an elaborate Gothic structure crowned by an ornate clock tower that still clangs the quarter hours throughout the day and night. Daily the young novice walked to and from the Brünn Theological College, where he attended classes in ecclesiastical law and archaeology, moral theology, exegesis, Hebrew, and Greek. And three times a day Mendel enjoyed big, bountiful meals, eating with such relish that by the time he was forty-five he was too fat to walk comfortably on the long specimen-collecting trips he had loved as a youth.

It was hard to resist those meals. The St. Thomas kitchen was widely known for its luscious special creations. Girls from the Moravian countryside would travel to Brünn to learn the culinary arts from some of the best cooks in the Hapsburg Empire. From there they would continue on to Vienna, hoping to achieve positions as servants to the Austrian aristocracy. In those days the highest praise for a Viennese household was to say that it featured "a Moravian kitchen."

The kitchen in which the girls most wanted to learn was the one at the monastery. Everything that emerged from that kitchen, presided over for more than thirty years by Chef Luise Ondrackova, was extraordinary. Especially prized was Chef Luise's famous rose-hip sauce. To make the sauce, which was used on meat, one uses jam made from rose hips, the brightly colored fruit of the rose. To a light white sauce, add wine. (Exact proportions were unheard of in the "kitchen books" of those days, one of which Chef Luise wrote after she retired.) Then, "for as many persons as you're preparing it, add as many teaspoons of rose hip jam to the sauce and add beef broth" (quantity of course unspecified). Then cook, stir, add salt and, "if necessary," vinegar.

But rose-hip sauce was only the beginning. Dinner, taken at midday, was always a three-course affair. A typical one in Chef Luise's day might include pea soup, pork with steamed green peas and boiled potatoes, and a rich strudel filled with chopped nuts.

Supper was more spare, with neither soup nor dessert; it might be a simple Moravian dish of, say, kidneys with brains and potatoes. But the monks never did without; each afternoon they were served a snack, which might consist of coffee, cream, marbleized black-and-white angel food cake, crescent rolls, cream tarts, and a selection of wines and liqueurs.

The refectory, the huge, cool hall where meals were taken, was the heart of the building and the heart of monastery life itself. The monks would gather at the long main table, fashioned of dark wood in the old German style, with Abbot Napp seated at one end and Prior Vorthey, presiding over his novices, at the other. As the church bell called the hours for breakfast, dinner, and supper, the eleven priests and assorted lay brothers and novitiates stood silently behind their chairs, listening as the refrains of the grace resounded from the vaulted ceiling. After the "Amen," the hall rang with a different kind of music — the scratch and rustle of chairs scraping the small floor tiles, the clatter of cutlery and crockery and glassware, and, beneath it all, the deep low murmur of a dozen or so men engaged in conversation.

There might have been some reading at the table, too, if Napp chose to conform to one of the rules written when the Augustinian sect was created in the fifth century. Seeing reading as the path to contemplation, seeing contemplation as the highest act of wisdom, and seeing wisdom as the goal of all Augustinians, St. Augustine encouraged his followers to read whenever possible. This often meant reading aloud from the Bible during meals. "It should be not only your jaws that are chewing food," one of the earliest rules of the order proclaims, "but your ears that are thirsting for God's word."

The monastery courtyard was connected, via a long avenue shaded by lime trees, to the lively Klosterplatz, a public square where fairs and carnivals were held. On festival days the Klosterplatz was a riotous mass of booths, umbrellas, sideshows, and merchandise, with people jostling each other to look at the goods or oddities on display. But once inside the monastery gate, quietness beckoned like a rumpled feather bed.

The monks' quarters were in a plain building with whitewashed stucco walls and a red tile roof, the windows small and unadorned. The courtyard and surrounding grounds were lush and reverberent with silence. If you walked through the monastery's side entrance, the informal way into the courtyard, you would first see the large vegetable garden. Beyond that, fruit trees dotted the lawn; farther still, a stone stairway was cut gracefully into the hillside, forging a path toward the higher meadows, where grapevines and wildflowers grew. On the right was the wing that housed the refectory on the first floor and the library above it; on the left, the local brewery, which peppered the air with a sharp, yeasty scent.

In Mendel's day the courtyard was also home to a tame fox, several squirrels and hedgehogs, and an assortment of birds, among them blackbirds, starlings, great tits, redstarts, and goldfinches. The brethren, nature lovers all, took great pains to protect these animals from predators. At one point they even enclosed part of the courtyard with mesh sheeting, in effect turning the monastery grounds into a giant birdcage.

As Mendel gardened in the cool of the morning, the birds kept him company — especially the fat-bellied blackbird with its neon-yellow beak, a solitary little soul, not unlike Mendel, singing a mellow song. That tendency for solitude, that persistence in repeating his single song again and again, would be called upon in years ahead to see Mendel through some terrible and disappointing times.

2

Southern Exposure

A garden is like the self. It has so many layers and winding paths, real or imagined, that it can never be known, completely, even by the most intimate of friends.

— *Deep in the Green*, Anne Raver

THE EIGHT-HUNDRED-YEAR-OLD pages rustle as you leaf through the Bible. They are made of vellum, more fine-textured than parchment because it is made from the stretched skin of young animals — calf, lamb, or kid — rather than from cow, sheep, or goat skin. The Bible's pages are dark at the edges, translucent when held to the light. They are smooth and almost white on what was the flesh side, suedelike on the reverse. The distinction is subtle but definite; the hair side tickles your fingers when you run a hand over it.

Like all twelfth-century books, this one is handwritten, in tiny, rounded calligraphy. The scribe began by making pinholes along the side of the page, then using graphite to delineate his margins and draw his lines. He began each page with red and blue filigree to highlight the first few letters of the first word. Corrections appear in the margins; like a teacher's markings on a child's composition book, they are outlined in red.

When the scribe finished his work, he would sign off with a colophon, a distinctive signature that might include a prayer of thanksgiving — or an expression of huge relief. Many medieval manuscripts ended with the scribe's complaints about the length of the book and a prayer for either eternal life, a jug of good wine, or the company of a pretty girl.

St. Augustine loved books like these. "If you pray," he once wrote, "you are talking to God; if you read, God is talking to you." Later, such books were required to be collected in the monasteries run by the order established in his name. Augustinian monasteries eventually became among the most significant repositories, after universities, of the knowledge contained in the written word.

The monks of Altbrünn took this mandate seriously. Their collection numbered 20,000 books, and they made sure the library was the most gorgeous room in the monastery. As St. Augustine had hoped, the library's very existence was a source of great comfort to men like Gregor Mendel, men who loved nothing more than to open a book and lose themselves in the cadences of great scholars, philosophers, and men of science. Here some of the monks, the ones with an intellectual rather than a devotional bent, were best able to hear God talking.

When Mendel came to the library, he would be dressed in his ordinary monastic attire: an ankle-length black soutane belted at the waist, with long, wide sleeves and a circular overgarment that fell, biblike, in a huge arc hanging to his belt, both front and back. The overgarment, an irksome extra bit of cloth, flapped up into Mendel's face at the slightest breeze. But the rest of the vestments were comfortable and kept the monks as warm as was possible in the drafty halls of their medieval living quarters.

Entering from the hallway, Mendel would pause at the steps leading up to the library to take off his black shoes, which more than likely were caked with mud tramped in from the garden. He reached into one of the cubbyholes stuffed with woolen slippers and put on a pair, as everyone was asked to do to protect the polished parquet library floors. Unlike the wide gray stones that paved the corridors or the tiny tiles in the refectory, the library's floor was a luscious wood that matched the elegance of the room. Hardwood bookshelves, their backs painted a vibrant blue, lined three sides of the enormous space. The fourth wall was composed entirely of five tall arched windows overlooking the courtyard, including views of the orchard, the glasshouse, and the brewery across the way.

This elegant room was not where Mendel and his fellow monks read their books. It was not even where most of the books were stored. All the books in this section of the library — and there were hundreds, maybe thousands — were for show, as was the room itself. Receptions were held here, and small dances, a grand piano in one corner of the room always available for musical counterpoint. But reading, study, contemplation? Straining to hear the very voice of God? Never.

To get to the true heart of the library, Mendel went to the southeast corner of the main room, to a bookcase with the same finely worked marquetry as all the others but with two shelves that were, curiously, empty. The half-empty bookcase rested on hinges, as did the cabinet below, and both swung out into the room to reveal a small doorway and five rough wooden steps. Mendel ducked his head — he did not have far to duck, for he was not a tall man — and walked down the steps. This put him in the first of four study rooms on the other side of the wall.

Here were the standing desks where Mendel and his more scholarly brethren did much of their work; over there, the old pine bookcases that held the less ornate reader's copies of the books in the collection. Catalogues helped the monks find what they needed; easy chairs were scattered about for them to occupy on languid, literate afternoons. The wall of windows continued in this secret sanctuary, running the full length of the library wing almost to the southern boundary of the monastery grounds.

There is a story, circulated long after Mendel died, that the monks at work in the library used to lean out the window and call to Mendel as he moved up and down his rows of experimental plants. Before the existence of this secret sanctuary became widely known, the assumption had been that the story referred to the windows in the formal reception room. But that was at the north end of the library wing. From the study rooms, where at least twice as many windows overlooked the courtyard to its southernmost tip, the monks would have been calling out toward a different spot altogether.

"What will it be for dinner tonight, Gregor?" they might have shouted down to Mendel, weeding between rows of carrots and cucumbers in the kitchen garden. "No peas tonight," he might have yelled back with a smile. "I'm saving them for something much more important than eating." His fellow brethren might have been surprised to hear Mendel admit that there was something more important to him than eating.

Why would it matter which windows they called from? Because the location of the windows helps us determine the location of the garden itself. And knowing where his garden was helps us know better how to judge the quiet monk. The solution to this old, and only recently solved, mystery has helped us discern the real Mendel from the Mendel trapped in myth. The solution resolved an apparent inconsistency in the monk's careful numbers that for many years made him seem less like an unappreciated genius than a liar.

Here is how the puzzle unfolded. If the monks were calling from the windows of the formal library, Mendel's garden would have been just below, in a narrow, fenced-in plot running along the library wing and separated from the more public sections of the courtyard by a long path, a fence, and a hedge. This is the garden shown in photographs, the garden that was analyzed and reanalyzed by generations of botanists, geneticists, and historians trying to figure out how Mendel could have done what he said he did in such a constrained space. This is also the garden that officials at the Mendel Museum, housed in the former refectory, still label as the site of Mendel's research.

But this plot was never big enough or sunny enough to have accommodated all of Mendel's experiments. It would have been too shady for peas; the strip of land was darkened by the shadow cast by the library wing itself during most of the morning hours. And it was so small and oddly shaped — sixty paces long by only twelve wide — that it is hard to imagine how, even when he had the added space of a greenhouse, Mendel could possibly have grown all the plants he said he grew.

Another part of the courtyard served nicely, however — a larger, sunnier, more versatile plot on the southern side near the service entrance gate. With its southern exposure, this plot was not in the shadow of any building until late in the afternoon. And its location fits better with the story of the brethren calling to Mendel from the library windows. They spent their time in the study rooms, not in the formal library, so they easily could have hailed their friend from there. And in order to hear them, he had to be working near the courtyard gate and not far from the greenhouse that was soon to be built.

The conventional wisdom about the location of Mendel's garden provided fuel for second-guessing his experiments through much of the late twentieth century. Because the garden near the monastery wall was too small for experiments on the scale he described, did he lie about how much crossbreeding he really did? And, if so, what else did he lie about? Because of the innuendo about whether the quiet priest fudged his data — rumors that persist, still without foundation, to this day — it was important to establish the exact location of his garden. The current thinking is that this bigger, brighter spot near the greenhouse, and not the one in all the photographs, was Mendel's plot — a location that allowed

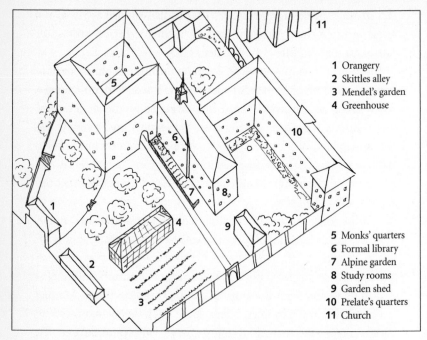

1 Orangery
2 Skittles alley
3 Mendel's garden
4 Greenhouse

5 Monks' quarters
6 Formal library
7 Alpine garden
8 Study rooms
9 Garden shed
10 Prelate's quarters
11 Church

The plan of the St. Thomas monastery in Brünn in Mendel's day.

him to grow his peas for years and years, unconstrained by space limitations or by the early disappearance of the sun.

The smaller garden alongside the monastery, though it might not be where he grew his peas, nonetheless held a special meaning for Mendel. Here he whiled away many hours during his first years at the monastery with Matouš Klácel (pronounced *klah*-tzell), a friar fourteen years his senior who would remain Mendel's closest friend for the rest of their lives. Klácel was a well-known philosopher and student of natural science. He had been in charge of the monastery's experimental garden since 1843 — the year Mendel arrived and the year the garden's previous custodian, Aurelius Thaler (last seen stumbling home drunk into Abbot Napp's angry arms), unexpectedly died. During the 1840s and 1850s Klácel grew alpine plants in this narrow strip of earth. He had transplanted them from the mountains of Moravia to the Brünn lowlands to test whether changes in the environment led to permanent changes in the plants themselves. He observed no changes. With this demonstration, Klácel helped repudiate the belief, common among biologists at the time, that plants and animals can metamorphose and pass on their new traits to their offspring in response to changes in living conditions.

Klácel was no doubt a radicalizing influence on his impressionable young friend. By the time Mendel met him in 1843, Klácel was in many ways a beaten man. Within a few years he would be relieved of his teaching duties as a philosophy professor in Brünn because of his writings in defense of *Naturphilosophie,* a German philosophy that combined evolutionary thought, a belief in purposeful activity in nature, and a view that the material world was a projection of a deeper spiritual reality. Among the most prominent German adherents of *Naturphilosophie,* which originated with the thinkers F. W. J. von Schelling and Lorenz Oken, were the poet Johann Wolfgang von Goethe and the philosopher G. W. F. Hegel.

Klácel's dismissal, in 1847, marked a low point in his confrontations with Church and government officials, who continually harassed him for his antimaterialist philosophy. That year Klácel wrote a three-part essay called "The Philosophy of Rational Good," which he submitted to the censors in Vienna, as required, before trying to publish it. But the censors, accountable as they were to the German-speaking Prince Metternich, concluded that the essay contained "harmful sentences" — no doubt the ones conveying Klácel's arguments in favor of Czech nationalism. He had used the *Naturphilosophie* view of evolution as an analogy to social evolution, applying a gardening image with his comment that "every age has much that is transient, which the dialectic sifts and polishes, until its kernel is revealed." In this way, Klácel implied, the "kernel" of the small Czech nation was likely to emerge from within the husk of German domination, and democracy was likely to emerge from feudalism — as long as Czech language and culture could germinate by being taught in the schools. These ideas were tantamount to revolution in the Hapsburg Empire, where the German-speaking minority was afraid of losing its grip on its feudal subjects.

A year later, in March 1848 — a year when revolutionary fervor was percolating through much of Europe, including Austria — Prince Metternich's government was indeed driven out. Klácel then took his ideas a step further, with such zeal that his friend Mendel went along with him, despite having few apparent political convictions of his own. Klácel drafted a radical petition stating that the priests of Moravia, isolated in their monastic lives, had been stripped of their civil rights and forced to live under conditions "representing the nadir of degradation." Chief among the petition's demands was the return to the priests of the full rights and privileges of citizenship, including the right to teach in the public forum. Seven priests altogether — including Klácel and Mendel — signed the petition, only to see it ignored and forgotten a few months later, when the emerging democracy's fledgling parliament was disbanded.

As political unrest roiled the continent, Mendel's life was undergoing a transformation, too. But in the end his sea change had nothing to do with the petition he signed, nor with political struggles at all. This was a purely personal, and humiliating, turning point for Mendel. As had happened many times before and would happen yet again, he arrived at this crossroads almost by accident, having traveled awhile in a particular direction mostly because a mentor had urged him to. And, as with the other crossroads, this one led him straight into a roadblock.

This time the roadblock concerned Mendel's inability to do the most basic work that priests were required to do. After five happy years at St. Thomas, Mendel was ordained and, too late, showed himself to be just about as unsuited for pastoral work as he had been for farming. In August 1848 he came down with a grave and mysterious illness. Its origin proved to be both surprising and, for Mendel, enormously embarrassing.

Mendel had been rushed through the steps of novice, subdeacon, deacon, and priest more rapidly than was customary — not necessarily because he showed any special talent for the job, although he might have, but because the monastery was running low on priests. Three young monks had died in 1847 alone, falling victim to whatever infectious diseases were killing the patients at St. Anne's Hospital, a short walk uphill along the winding Bäckergasse, where the monks had been asked to administer last rites.

So Mendel was ordained on August 6, 1847, just fifteen days after he turned twenty-five, the minimum age for a priest. He spent the next year completing his theological studies, and in August 1848 assumed some of the chores of a parish priest, ministering to the sick, the dying, the impoverished, the infirm. By early the following year, it had all proved too much for him. Mendel took to his bed once again — and did not rise for more than a month. Unlike his fellow priests, however, he had not caught one of the infections of his parishioners. No, his disease was emotional — though it was as real as if he had contracted tuberculosis or typhoid fever.

Lying in bed in the first weeks of 1849, suffering both physically and spiritually, Mendel was rescued once again, this time by his ever-indulgent abbot. Napp could plainly see that Mendel would make a terrible priest. Perhaps, he thought, the young man should try teaching instead.

"He is very diligent in the study of the sciences," the abbot wrote in mid-1849 to Bishop Schaffgotsch, but "much less fitted for work as a parish priest, the reason being that he is seized by an unconquerable timidity when he has to visit a sick-bed or to see any one ill or in pain. Indeed, this infirmity of his has made him danger-ously ill."

Schaffgotsch was not especially inclined to do Mendel any fa-vors. Even though this was years before their confrontation over Mendel's experimental mice, the bishop found the younger man's waggish sense of humor not the least bit funny. During his novi-tiate days, Mendel had muttered to a fellow monk in what he thought was a whisper that the closed-minded bishop was obvi-ously possessed of "more fat than understanding." Schaffgotsch never forgot the sting of overhearing that comment — especially coming from a rotund fellow like Mendel.

But Napp insisted that the young priest be given dispensation to teach, for this was the only way to guarantee Mendel a life he could tolerate. The timing was propitious: there was, after all, an edict from the Austrian authorities requiring local clerics to give some-thing back to the communities in which they lived, and teaching in the secular schools was considered one of the best ways to do this. So Schaffgotsch finally relented, agreeing to send young Mendel to the ancient town of Znaim (called Znojmo in Czech) in southern Moravia to try his hand at teaching in the Gymnasium.

In the fall term of 1849 the twenty-seven-year-old Mendel taught elementary mathematics and Greek to the boys in the third and fourth forms. He was judged by his fellow teachers to have "a vivid and lucid method of teaching," employing "zeal and tenac-ity," despite his inexperience and lack of formal training. And ac-cording to the police reports that monitored just about everyone's comings and goings in those revolutionary days, Mendel com-ported himself in a manner entirely appropriate to a man of the cloth. Yes, he went to the theater six times during his year in Znaim, "but always in the society of one of his colleagues."

In a closely controlled police state where everyone was account-able to the authorities, Mendel still managed to study and teach the natural sciences, propelled by sincere and innocent curiosity and by a love of both knowledge and the natural world. But natural scientists, if they are intellectually honest, often find themselves taking heretical positions on matters of creation and procreation, positions that challenge the very underpinnings of the Catholic Church. The issues that concerned Mendel, such as reproduction, heredity, and the continuity and discontinuity of all forms of life, would prove to be the fulcrum for the most impassioned debate of the nineteenth century: the debate over evolution. In the middle of the century, before Charles Darwin's revolutionary book was published, the controversy tended to focus on the one issue that represented the two competing halves of Mendel's own existence: the inevitable animosity between science and God.

3

Between Science and God

> *Be fruitful, and multiply, and replenish the earth,*
> *and subdue it.*
>
> — Genesis 1:28

IN THE BRICK ENTRANCEWAY of the town hall of Brünn — which, since its incorporation into Czechoslovakia in 1918, has been known by its Czech name, Brno (pronounced *burrr*-no) — there dangles from the ceiling a grinning stuffed alligator. Tradition has it that this animal, the so-called Brno dragon, was tricked into eating a carcass filled with lime and, to slake its unbearable thirst, went down to the River Svratka (Schwarza in German) and drank until its belly burst. More likely, but less interesting, is that the alligator was a gift to the city from Archduke Matthias, who in turn had received it from a Turkish sultan sometime in the late sixteenth or early seventeenth century.

Despite its snaggle-toothed grin, the alligator looks no more threatening than a plastic piñata — albeit a rather grisly piñata — hanging flat-footed over the musty entrance. But it brings to mind not only the eager gullibility of the people of Brno, but their remoteness from any direct understanding of the natural world. What, after all, is a creature from the Amazon doing in a medieval town in the middle of Europe?

At the time Brno received its "dragon," the goal of natural historians was to describe and categorize reality. And the compelling force behind this urge to categorize was to uncover the beautiful, orderly, prescient, and perfectly clear-sighted design of the Creator. Natural historians fancied their work as not all that different from the task of theologians. They believed that by achieving a richer understanding of nature they would have a window into the mind of God. Indeed, the two interests, science and religion, often resided in the same individual. Charles Darwin, arguably the most brilliant naturalist of the nineteenth century, set out at first to be a parson. "The pursuit of Natural History, though certainly not professional, is very suitable to a clergyman," wrote Darwin's uncle Josiah Wedgwood, trying to reassure Charles's father about the son's career goals. Clerics frequently dabbled in the sciences.

Then, in the middle of the nineteenth century, God died. His death was announced most dramatically and famously by the German philosopher Friedrich Nietzsche, whose obituary for God in 1883 helped create the new branch of atheistic philosophy known as existentialism. However, the death of God and the ascendancy of science had happened with a cyclical regularity ever since the days of Jesus Christ.

One of the more dramatic confrontations between science and God took place in the sixteenth and seventeenth centuries, in the long struggle over the heliocentric view of the universe — the notion that the earth revolves around the sun rather than the other way around. In 1543, the year of his death, Polish astronomer Nicolaus Copernicus published his revolutionary ideas. But in those days, heliocentrism was not considered heretical; in publishing his book, Copernicus had the blessings of Pope Paul III.

Seventy-three years later, in 1616, Galileo Galilei, an Italian mathematics professor and devout Catholic, faced excommunication for defending those same ideas. Official doctrine had changed by then, and the Church considered the Copernican view to be "false and altogether opposed to Holy Scripture." But Galileo, a brilliant physicist, astronomer, and inventor as well as a mathematician, refused to renounce his radical ideas before the Inquisition.

The Church's job, he supposedly said, is to tell people how to go to heaven, not how the heavens go. In the end, Galileo was placed under house arrest from 1633 until his death in 1642.

When Church doctrine once more loosened its grip on scientific inquiry, the work of the scientist changed. Indeed, the very designation "scientist" reflected a dramatic change; until the early 1800s, men like Darwin were known as naturalists or natural philosophers. But in 1833 Cambridge philosophy professor William Whewell proposed that members of the British Association for the Advancement of Science should be called "scientists." In a stroke, Whewell professionalized a group that until that point had been a close brotherhood of amateurs. Around this time the newly named "scientists" no longer viewed their task as the gradual peeling away of evidence of God's omniscience. Now they believed their job was more straightforward but also more difficult. They saw the elucidation of the laws of nature not as a divine plan to be uncovered, but as a secular puzzle to be solved.

One mechanism for uncovering the divine plan for life had been to collect its most outlandish examples, which led to a phenomenon of the sixteenth century called Cabinets of Curiosities. A second approach emphasized similarities among forms of life rather than oddities, leading to the branch of science called taxonomy.

A Cabinet of Curiosities was a large room lined with shelves housing collections of fascinating objects from the natural world, objects with a hint of the macabre, a sort of real-life wax museum that was the early forerunner of contemporary museums of natural history. The objects were hung from the ceiling, piled up along the walls, crammed into arched passageways. Such a room might have housed the Brno dragon, since many Cabinets of Curiosities included at least one stuffed alligator. The bigger and more bizarre these collections were, the better. The Italian naturalist Serpetro, for instance, became famous in the seventeenth century for a "wonder room" that included mandrake roots — thought to have magical properties because of their humanlike shape — petrified objects, giant's bones, stuffed Pygmies, and "fabulous zoophytes" such as Scythian lambs. Collectors displayed their preserved specimens dramatically, constructing "gruesome tableaux" by injecting the arteries with wax to stiffen the carcasses, allowing them to be arranged in the most morbid poses. Some created fanciful hybrids of their own, attaching a stuffed top half of a child's body to a stuffed lower half of a fish to create a weird "merbaby." Many Cabinets of Curiosities featured stuffed birds of paradise, usually footless, because the birds' feet were almost invariably injured during capture — thus inadvertently spreading the tale that birds of paradise had no feet and, once airborne, could never land.

A different approach to finding the divine order in nature was through the systematic arrangement of the tree of life, the field of study known as taxonomy. The leading taxonomist of his day — indeed, of any day — was Karl Linné, a Swedish botanist who did his most famous work in the 1730s. Writing under the Latinized version of his name, Carolus Linnaeus, he constructed a system of categorizing all living things that is still in use today. Linnaeus divided organisms into two kingdoms — plant and animal — and further divided the kingdoms into classes, orders, genera, and species. Basing his plan as far as possible on structural similarities, he categorized every creature on earth; from what had seemed to be disorganized chaos, clarity arose.

In Linnaeus's view, each of these subdivisions had a particular source. The "Creator of the Universe" clothed the different members of his two kingdoms in "diverse constituent possibilities," giving rise to the orders, he wrote in *Species Plantarum* in 1753. From the orders — he seems to have skipped over classes here — the "Omnipotent" created genera. "Nature" blended the genera into species, and finally "chance" blended the species into the smallest of the Linnaean subdivisions, varieties.

Under the Linnaean system one can classify a fuchsia, Mendel's favorite flower, in a way that shows at once what its nearest relatives are. The fuchsia belongs to the kingdom Plantae, class Angiospermae, order Myrtales. It is related to all other Plantae, but most closely to other Angiospermae (flowering plants) and even more closely to Myrtales. There are about a hundred species of fuchsia, each with its own Latin binomial (two-name) designation. One of the most common is known scientifically as *Fuchsia magellanica*. *Fuchsia* is the genus — which is always, according to the conventions of the Linnaean system, capitalized and italicized — and *magellanica* is the species — always in italics and lowercase letters. The species designation is unique and specific; the very word "specific" derives from "species." *Fuchsia magellanica* is a name given to no other angiosperm, no other Myrtales. When you call something *Fuchsia magellanica*, horticulturists around the world, no matter what their native language or climate or gardening interest, will know exactly which plant you mean.

Similarly, modern human beings (kingdom Animalia, order Primates) can be designated by genus and species, respectively, as *Homo sapiens*. Early man belonged to other genera — *Australopithecus afarensis*, *Paranthropus boisei* — and, more recently, to other species within our own genus — *Homo erectus*, *Homo neanderthalensis*. But only humans of the past hundred thousand years or so go by the Linnaean binomial *Homo sapiens*.

Since Linnaeus's time, naturalists have had to add new categories to accommodate the discovery of more and more species, so today the taxonomy of living things is divided, in descending order, into kingdom (of which there are now five rather than two), phylum (or, in plants, division), class, subclass, order, family, genus, and species — or, to go back to the fuchsia again, Plantae, Magnoliaphyta, Magnoliopsida, Rosidae, Myrtales, Onagraceae, *Fuchsia magellanica*.

Linnaeus became a national hero in Sweden, but while he

boasted about his accomplishments in writing, some have read anguish rather than arrogance between the lines. "God has bestowed on him the greatest insight into nature-study," he once wrote about himself, using the third-person form common in autobiographies of the time. "None before him has written more works, more correctly, more methodically, from his own experience. None before him has so totally reformed a whole science and made a new epoch." But he also, rather modestly, considered his scientific acumen to be proof of God's beneficence. "God has permitted him to see more of His created work than any mortal before him." And for all that, for all his special stature on this earth, even he, like every other creature, was destined to suffer and die.

Linnaeus was a genius, whose road toward insight was probably somewhere between the 99-percent-perspiration route and that of sheer, playful intuition. But he was said to be a personally unpleasant man: haughty, egocentric, moody, and often difficult to bear. Mendel was not a genius of the brooding kind, like Linnaeus, except when he took to his bed in despondency. But he was not especially easy to live with, either. Mendel could be so intense, so single-minded — and at times so wrong — that the patience of even his most loyal friends and associates occasionally wore thin.

Linnaeus did help breathe life into an entirely new scientific epoch — the Age of Enlightenment. His motivation was one of profound devotion. "God created; Linnaeus arranged," went the adage, and while Linnaeus enjoyed the fact that his name and God's were uttered in the same sentence, he was not entirely without humility. He never once questioned the primacy of God's place in that hierarchy.

Through most of the eighteenth and nineteenth centuries, scientists believed that the adaptations of every organism on earth were evidence of a divine plan, a plan that, for most, ended and began with man. People tended to indulge in a human-oriented view of life. Many believed, for instance, that horses' backs were shaped simply to give man a comfortable seat for riding; that seawater contained alkaline substances like magnesia and lime so sailors could clean their clothes without using soap; that tides existed just to help the ships come in.

This was a comforting view of the universe. Everything was orderly, everything was preordained, and everything existed for the aid and benefit of mankind. This view was seriously challenged in the middle of the nineteenth century, when the physicist Rudolf Clausius discovered entropy. Nature was not comforting and orderly at all, he said; objects tended to move in the direction of extreme disorder. The second law of thermodynamics, which Clausius formulated, also called the law of entropy, stated that every isolated system becomes more disordered over time. In this construction an orderly system is as regulated as a crystal, with each atom in precisely the right spot, while a disordered system is like vegetable soup, arranged at random without any clear pattern or organization. And the flow of nature is toward greater randomness, patternlessness, and disorganization — an unsettling view for people who saw God's logic in every flutter of a butterfly.

The year 1850, when Clausius discovered entropy, was the same year that a completely different kind of disorder came into the quiet, well-managed life of Gregor Mendel. In the summer of that fateful year, Mendel encountered the biggest challenge he had yet confronted, and in its face he crumbled. His shameful performance would throw him into despair and cause him to wonder whether he had anything to offer the world, either then or in the future.

4

Breakdown in Vienna

A garden is always a series of losses set against a few triumphs, like life itself.

— *At Seventy,* May Sarton, 1912–1995

No ONE WAS PLEASED about being in the stuffy examination room on that August afternoon. Vienna was hot and uncomfortable, and the esteemed professors chafed at being forced to remain in the city during their two-month summer holiday. How they longed to be somewhere else — preferably with their families in the cooler, greener countryside. But a series of missed communications had led to this moment, and now in a university classroom sat six bothered men. And there stood Gregor Mendel, sweating.

He had argued his way into this classroom, and now he regretted having done so.

The date was August 16, 1850, a Tuesday. The six men, current or former professors at the University of Vienna — Professors Bonitz, Enk, Grauert, Kner, Lott, and the chairman, Count Andreas von Baumgartner — were sitting in judgment of Mendel's qualifications as a high school science teacher. The men's irritation showed. Their impatient manner made Mendel stumble as he struggled to answer their questions, most of which related to physics. They had come into the room with already low opinions of Mendel, based on his pitiful performance in the written examination the day before. Now his performance in the oral examination, the *viva voce,* was turning out to be hardly more impressive.

A rare moment of bravado had led Mendel into this hornet's nest. The previous week he had arrived, full of expectations, in the government offices of chairman Baumgartner, who was also a renowned physicist and the minister for public works. In a letter dated August 1, Mendel had been told to report to Vienna for his written and *viva voce* examinations. These constituted the second and third parts of the teacher certification process, which had begun in May with two essays Mendel had had to write just as the school year in Znaim was ending. Those essays, concerning meteorology and geology, had been judged by the board of examiners to be not especially remarkable, but acceptable enough to allow Mendel to move on to parts two and three.

But soon after the August 1 letter, Baumgartner had sent a second one with different instructions, telling Mendel not to come to Vienna after all. The members of the committee, Baumgartner wrote, wanted to delay the exams until fall, so they could leave for their summer holidays on August 12.

Mendel never received the second letter. If it did arrive, it came after he was already aboard the train to Vienna. The journey to the capital took nearly half a day, and in his innocence Mendel spent most of the time full of hope and expectation. In retrospect, given the dire consequences at the end of the journey, that half-day of travel was one he would gladly have undone, one he mentally replayed over and over again in reverse, so he would never have to arrive in Vienna at all.

"Forgive me, Herr Minister, but I did not know you had changed your preferences regarding the timing of my arrival in Vienna," Mendel must have said — respectfully, though perhaps seething underneath — when Baumgartner told him to go home. "The last communication I received from you was sent on August 1, and instructed me to come to Vienna and report directly to your office. Now that I am here, it is very important for me to continue, Herr Minister. I would very much like to receive my teacher's certificate before the beginning of the fall semester."

Did Baumgartner take that opportunity to warn Mendel about the state of mind of his examiners? Did he tell him that trying to assemble the others at this inconvenient time might make them especially churlish and inclined to find fault with Mendel's performance? If he did, the priest was undaunted. He was there, he was ready, and he wanted to begin.

Such confidence was rare for Mendel. The previous school year had been such a good one for him that the summer of 1850 found him in unusually high spirits. He loved teaching at the Znaim Gymnasium; he could see, as could his colleagues and students, that he had a gift for pedagogy. He already entertained fantasies of a rewarding teaching career, adored by his students, respected by his fellow teachers, fulfilled in his work.

The poor timing of his August appearance was not the only roadblock to Mendel's quest for certification. He had a severe case of test anxiety, and his first two examination essays, written at home before his arrival in Vienna, were little more than mediocre — or worse. Baumgartner had found the first one, on meteorology, to be "satisfactory" though not remarkable. But Professor Kner, judging the geological essay, was distinctly unimpressed. He used words like "arid, obscure, and hazy" to describe Mendel's thought process; "erroneous" to describe his facts; "hyperbolic" and "inappropriate" to describe his writing style.

Mendel believed he would perform better in Vienna. But he was wrong.

Baumgartner eventually relented and agreed to have Mendel's trials begin on August 15. That is when things started to fall apart for the young priest. The result was a failure so spectacular — the way a train wreck is spectacular — that it would take Mendel years to recover from the ignominy.

Subsequent teaching evaluations showed Mendel to be a bright man with an incisive, highly organized intelligence and a clear writing style. But the written examination essays of 1850, especially the one in zoology, bring to mind a man in agony, reaching into the dark recesses of his memory, pulling up only the lint in an empty pocket. It was as though Mendel were breaking apart, physically and mentally, before his examiners' very eyes.

In the zoology essay he was asked to classify the mammals and to mention some of their uses to man. He mentioned six orders of animals, failing to use the proper scientific terminology for a single one, instead calling them by idiomatic terms such as "quadrumana," "quadrupeds," "plantigrades," "clawed ungulates," "hoofed ungulates," and "web-footed animals etc." To the question of how they were useful to humans, his answer sounded like the throat clearing of a high school student who has not done his homework and is hoping to get away with it. Among the useful clawed ungulates, he wrote, were "the cat, a useful animal because it exterminates mice, and because its soft and beautiful fur can be dressed by furriers" and "the civet, whose anal glands secrete an aromatic substance which is an article of commerce." Regarding useful hoofed ungulates, his answer was even more strained. He simply listed:

> the horse;
> the ass;
> the ox;
> the sheep;
> the goat;
> the chamois, the deer, and the stag;
> the llama, much used in Mexico as a beast of burden carrying light
>   loads up to one or two hundredweight;

the musk ox;

the reindeer;

what the reindeer is for the north, the camel is for the hot steppes;

the pig;

the elephant is a splendid beast of burden.

Not a formal designation of genus or species in the lot; not a Latin name among them; not a sense, anywhere, that Mendel was doing much more than free-associating. Even Hugo Iltis, who later taught biology in Brünn at the same school as Mendel, and whose reverential *Life of Mendel* was to become the standard biography for more than sixty years, thought this answer mystifying and "at times positively absurd."

The examiners would have agreed with this harsh but accurate assessment. Kner, who was already biased against Mendel because of his poor showing in the geology essay — and because Mendel had foolishly failed to cite, or even to read, Kner's own textbook of zoology — was especially ruthless. "The part of the question which relates to the utility of animals as yielding materials for industrial purposes or for use in medicine is answered in the most schoolboy fashion," he wrote. "The candidate seems to know nothing about technical terminology, naming all the animals to which he refers in colloquial German, and avoiding systemic [Linnaean] nomenclature. . . . This written examination paper would hardly allow us to regard him as competent to become an instructor even in a lower school."

But in the *viva voce*, the examiners agreed that Mendel showed, if no special brilliance or acumen, "unmistakable good will." All, Kner included, were willing to give Mendel the benefit of the doubt. The young man is "devoid neither of industry nor of talent," Kner admitted, adding that Mendel's biggest problem was that he was largely self-taught. "Still we may hope," Kner wrote in an opinion separate from the other five examiners' joint report, "that if he is given opportunity for more exhaustive study together with access to better sources of information he will soon be able to fit himself."

With this in mind, Abbot Napp granted Mendel a great favor, probably the most decisive favor of his life. He agreed to send him to the Royal Imperial University of Vienna to study and, as Kner put it, "to fit himself."

Church bells in Vienna ring sweetly musical, nothing like the off-key metallic thud of the Augustinian chapel bells in Brünn. In both cities today, automatic timers toll the quarter-hour all night long, taunting the insomniac in dreadful increments. But in Mendel's day, the bells were rung by hand. If he found himself unable to sleep, either in Brünn or during his two years as a student in Vienna, his nighttime stirrings would have been accompanied by silence — until about four o'clock on a spring or summer morning, when skies dawned an electric blue and the birds began to sing.

Mendel had much to keep him awake in either city. In Brünn he could go over ideas for his mouse breeding experiments as the rodents scratched in cages just a few feet from where he lay. In Vienna he had his schooling to fret over, since he was carrying twice the ordinary class load and filling himself to the brim with the latest thinking in all the natural sciences of his day. But if he was wakeful at night, there was no hint that he was ever tired or distracted during his energetic days.

Vienna was the apogee of Mendel's transformation from Silesian peasant to educated natural scientist. He had as much freedom to taste new things as a hungry boy in a candy shop with coins jangling in his fist. But even though these two frenetic years would, as we can see in hindsight, pave the way for his most glorious research, vindicating a life spattered with troubling failures, his sojourn started off poorly, for he was forced to arrive a month after classes had begun.

The redoubtable Bishop Schaffgotsch of Brünn gave Mendel permission to venture to Vienna only on condition that he "lead the life proper to a member of a religious order," by which he meant residing in a monastery or rectory, and "not become estranged from his profession." But despite Abbot Napp's entreaties, no Viennese monastery was able to house Mendel for the duration of his studies. The Brothers of Mercy told Napp they had no room to spare, as their three guest rooms were always occupied and the monastery's own brethren were forced to double up.

Napp sent Mendel to the capital anyway. He could envision no other way for the young man to obtain the university training he so clearly craved — and, if he were to become a successful schoolteacher, so obviously needed — than to take the chance that the Bishop might not wholeheartedly approve of the living quarters he found. So on October 27, 1851, Napp dispatched the unschooled and untraveled monk on the night train to Vienna to fend for himself. This was a risky decision, since Vienna, despite its reputation as a festival of scientific and cultural confections, allowed for such brazen pursuit of pleasure that one of its favorite sons, the playwright Franz Grillparzer, called it "the Capua of the mind." It was in the wealthy Italian town of Capua that Hannibal's troops had abandoned themselves to reckless overindulgence, rendering them unfit for further warmongering and eventually bringing an end to the Second Punic War. The same dangerous overindulgence, Grillparzer believed, could trap the unwary in nineteenth-century Vienna, too.

After searching through the unfamiliar city for days, Mendel finally secured housing in a large corner apartment building in Vienna's third district. The lodging, at 358 Landstrasse, near the convent of St. Elizabeth's, was a long walk from the university, on the other side of the Vienna River, at the time a disease-ridden underground waterway that had not yet been channeled into the Wienfluss Canal. At the site of today's Stadtpark was a mineral-water pavilion, the source even today of Vienna's drinking water, reputed to be among the best tasting on earth.

After settling in, the priest found himself once more in the offices of Baumgartner, by this time Austria's minister of trade. Mendel arrived on November 5, 1851; classes had begun on October 5. He presented Napp's letter of introduction: please allow the bearer of this letter to commence his studies at university today, even though he is quite tardy. Otherwise he will be forced to wait until winter semester, and his progress will be further delayed. Baumgartner, who remembered Mendel from his exam failure the year before, remembered also Mendel's "unmistakable good will" and allowed the young man to begin.

Mendel tried to make up for lost time by taking on a grueling schedule; beginning in his second semester, he spent thirty-two periods a week in lectures and practicals, when the normal load was twenty. Although at twenty-nine he was seven or eight years older than most of the other students, Mendel always felt he was a month behind everybody else.

Almost as soon as he arrived, Mendel became an assistant demonstrator at the Physical Institute. This was a position reserved for the brightest and best, the most serious of physics students. There were twelve slots for such demonstrators, called *élèves* (French for "students"), all of whom were required to be secondary school teachers. By the time Mendel showed up in November, all twelve positions had been filled. But something about Mendel's promise, or enthusiasm, or persistence impressed the famous physicist Christian Doppler, the director of the Physical Institute, whose health was failing rapidly. Doppler agreed to bring the priest on board as *élève* number thirteen.

Doppler, a clear but dull lecturer, taught the course in experimental physics in which Mendel was enrolled. How wonderful it would have been if he had the flair of another physicist, Christoph

Buys Ballot of Holland, for demonstrating his own most famous discovery. Buys Ballot, in Utrecht, put a band of trumpeters on an open car and attached the car to a locomotive. As the bizarre setup moved and the band began to play, observers at one end of the train track could hear the change in pitch of the trumpets' blare, even though the musicians were playing a single, sustained note. Buys Ballot thus demonstrated, with far more fanfare than Doppler ever did in his own classroom, the perceived change in the frequency of sound waves from a moving object, which Doppler had discovered and which we now know as the Doppler effect.

Doppler was ailing from a chronic lung disease contracted when he taught in Prague ten years earlier. He took a medical leave of absence in 1851 and retired in 1852, moving to Venice in search of an easier climate. He died there the following year, at the age of forty-nine. By then, Doppler's teaching responsibilities and the directorship of the Physical Institute had been taken over by an equally brilliant physicist, a man who was probably Mendel's most influential teacher: Andreas von Ettingshausen.

Ettingshausen was best known for his *Combinationslehre* (combination theory), which he had developed to describe the relationship among the objects in a group arranged in any predetermined way. It was a mathematical way of determining all the possible arrangements of any group of things, whether natural or manmade — people, ants, numbers, sentences, colors, peas. Later it would occur to Mendel that combination theory might also govern the arrangements of hereditary determinants. This was a fruitful line of inquiry, but some of Mendel's other attempts to use combination theory turned out to be dead ends. He was, in a way, like the child with the hammer, to whom every problem starts to look like a nail. He had a tool with which to measure and clarify the world, and he was determined to make every situation he confronted bend to the tool's special abilities. This would prove quite useless in some situations, as when he tried, in the final years of his life, to apply the theory to the long compound words that formed common German surnames.

His belief in combination theory was a mark of Mendel's genius. Throughout history, some of the most creative minds have been those capable of maintaining two different mental constructs of the world simultaneously and applying the principles of one model to problems in the domain of the second. As Mendel worked on questions of inheritance, he used basic mathematical principles to pose questions in an entirely new context. He was instinctively demonstrating the dictum issued years earlier by the French philosopher Souriau: *pour inventer il faut penser à côté* — "to invent, you must think aside" — that is, slightly askew.

At the same time that Mendel was a synthesizer of ideas that did not obviously fit together, he was also a product of his day. The mid-nineteenth century was characterized by what one historian called an "avalanche of numbers." All over Europe, scientists were counting things: the number of killings with knives, the distribution of heights of soldiers, the causes of death in a population. Not every quantitative approach involved combination theory, but the frenzy of numbers showed that scientists everywhere, in every field, were ready to try new routes to making sense of the natural order of things.

The influence of mathematics in general, and Ettingshausen in particular, would help Mendel negotiate the maze of data he amassed from his pea experiments. But the guiding spirit behind those experiments was not so much Ettingshausen as another of Mendel's professors in Vienna: Franz Unger, who taught physiological botany. It was through Unger, one of the most controversial biologists of the time, that Mendel first learned about the landmark hybridization work of his most famous predecessors, the

Germans Josef Kölreuter and Karl Friedrich von Gärtner. Both men developed breeding methods that Mendel would soon bring to life again in his garden. Unger also introduced Mendel to the revolutionary new approach to "scientific botany" then being promoted by Matthias Jakob Schleiden, the codiscoverer (with Theodor Schwann) of the theory that all plants and animals are composed of cells. Schleiden considered biology as rigorous a science as physics and chemistry, with a need for a new, more deductive experimental approach in the search for generalizable laws.

During his botany lectures, Unger spoke approvingly of Schleiden and his cell theory and described earlier hybridization experiments, many of them from Great Britain, that had involved a plant particularly well suited for crossbreeding because of the structure of the flowers and because of its easily identifiable character traits. The plant was the common, or garden, pea, the species *Pisum sativum.*

Much less providentially, it was also through Unger that Mendel developed a respect bordering on reverence for Karl von Nägeli. Unger liked to say that Nägeli, a professor of botany at the University of Munich, had provided the blueprint for the study of plant physiology when, in 1842, he described, with stunning accuracy, the processes of cell division and seed formation in flowering plants. Unger's reverent attitude may have been warranted, but Nägeli would ultimately prove to be a source of Mendel's undoing.

Like Nägeli, Unger took a mechanistic, "hard science" view of botany. The task of the physiological botanist, as Unger put it, was "to reduce the phenomena of life to physical and chemical laws." As for the most controversial issue of the day — the fixity or nonfixity of species — Unger took a characteristically heterodox position. He believed that a process of metamorphosis was somehow involved in "transmutation," the word then generally used to mean evolution. But he did not believe that metamorphosis could by itself explain the diversity of species on earth. Nor did he venture to guess how transmutation might occur or what mechanism might underlie it.

Sitting alongside Mendel in Unger's lecture hall were several young men whose lives would intersect his again and again. One was Anton Kerner von Marilaun, who would go on to become a highly regarded botanist — and to whom Mendel later sent a reprint of his paper on peas.

Another was Johann Nave, a law student at the university who also attended lectures in the natural sciences. Nave and Mendel sat in on Unger's lectures together — and they continued to discuss botanical matters back in Brünn, to which Nave moved in 1854 to enter the civil service. Along with Mendel, Nave was a founding member of the Brünn Society for the Study of Natural Science, and he developed a reputation as the world's leading expert on the algae of Moravia. He might have been a role model for Mendel, proof that it was possible to spend one's days pursuing a profession — in Nave's case, the law — and still become proficient enough in the natural sciences to make a true contribution to the steady accretion of knowledge. Nave was also, until his early death in 1864, one of Mendel's closest — indeed, one of his only — friends.

Mendel returned to the monastery in Brünn in July 1853, when his two years in Vienna came to an end. Much of his time is unaccounted for over the next year, but soon he fell into a pattern of study, teaching at the Realschule, and gardening. At the monastery he spent long hours in the glasshouse, breeding peas to be sure they did in fact breed true.

The summer his pea breeding began, 1854, would turn out to be Mendel's last season in the glasshouse. Napp had greater plans for him, embodied in part by the greenhouse he hoped to build on Mendel's behalf. The abbot was collecting permits and architectural drawings for a structure that would be big enough to house all the plants Mendel needed to reach some meaningful conclusions. Although Napp did not fully understand the mathematics, he understood one inescapable statistical truth: the greater the number of examples undertaken, the more reliable the results. As Mendel would later explain, large numbers are necessary, "because with a smaller number of experimental plants . . . very considerable fluctuations may occur." To deduce "true numerical ratios," he said, requires "the greatest possible number of individual values; and the greater the number of these the more effectively will mere chance be eliminated."

Because of the arrangement of the courtyard, there was only one spot for a greenhouse that would be big enough. Over the next two years Napp saw to the building of it, so that Mendel's experiments, which would prove to be the source of his greatest heartache and his ultimate triumph, could at last get under way.

As this was about to happen, Mendel, happily hurtling down the road of life, was stymied by another breakdown — this one even more devastating than the one before.

5

Back to the Garden

> *A garden is a grand teacher. It teaches patience and careful watchfulness; it teaches industry and thrift; above all it teaches entire trust.*
>
> — *On Gardening*, Gertrude Jekyll, 1843–1932

IN MENDEL'S TIME it took a full morning to get from Brünn to Vienna; although less than ninety miles apart, the cities were separated by four hours on the train. The scenery glided past at a leisurely fifteen or twenty miles an hour as the train moved south, passing landmarks arising from flat countryside dotted with fields of maize. Here was Raigern, with its eleventh-century Benedictine abbey; there the park of Prince Liechtenstein, its elaborate summerhouse rising two hundred feet to provide the best view of the nearby town of Saitz. At nearly the halfway point, the tracks crossed the River Thaya, which formed the border between Moravia and Austria. The Little Carpathian Mountains rose to the east, the Leopoldsburg castle to the west. Almost at journey's end were the wooded islands of the Danube, which the train crossed via a half-mile-long iron bridge. With the tower of St. Stephen's Cathedral looming in the distance, the ride ended at Vienna's North Station.

Mendel decided to set off on this journey one more time in the spring of 1856. He wanted another try at passing the certification examination he had botched so badly six years before. Having spent the previous two years as a full-time substitute teacher at Brünn's new Realschule (the more modern, technological equivalent of a Gymnasium), he believed the time was finally right.

So much, after all, had happened between the doomed summer of 1850 and the promising spring of 1856. After the two transforming years at the University of Vienna and four productive years teaching, reading, and beginning his experiments in the garden, he had developed a deeper understanding of the basic laws of algebra, statistics, and combination theory. He had a surer footing in the botanical sciences that most compelled him and a more thorough knowledge of the aspects of natural science that were most important for teachers to understand. How could his performance do anything but improve the second time around?

But apparently not enough had changed. After just one question on his *viva voce* examination, Mendel gave up — and failed once again.

Both failures were almost certainly the result of his debilitating test anxiety. The second one, though, seemed far more extreme than the first. Having stumbled while answering the first question, it seems that Mendel simply chose not to go on. But no one can say for sure what really happened that spring day in Vienna. Relegated for the rest of his career to the rank of uncertified substitute teacher, Mendel never once talked about the details of his second test fiasco.

In the face of his silence, rumors bloomed. One of the most interesting seems to have originated with his fellow monks, who liked Mendel and could not bear to think of him as merely inept. In their rendition, Mendel failed not for lack of courage but for its excess. He failed because his nerves were too steely, his spine too strong, his integrity and intellectual honesty simply too uncompromising for his own good.

According to this version, Mendel had a confrontation with one of the examiners, Eduard Fenzl, director of the Vienna Botanical

Gardens and one of Mendel's professors at university. Fenzl was a "spermist," meaning he believed the plant embryo resided, microscopic but entirely preformed, in the pollen, and passed to the ovary through the pollen tube. All it needed to do was grow; the female part of the plant offered nothing more than an environment that made growth possible. Mendel, in contrast, believed that both male and female gametes contributed equally to the offspring's makeup. In his view embryos were not preformed but were created anew with every fertilization. The two men supposedly argued this point, and Mendel stubbornly — and characteristically — refused to back down, choosing failure over capitulation.

The story continues that it was to resolve this battle that the priest began his *Pisum* experiments. If that is true, this tale ends with a nice little twist: of the three scientists credited with rediscovering Mendel's work at the dawn of the twentieth century, with finally bringing it out of the musty recesses and onto center stage of the nascent field of genetics, one turned out to be Fenzl's own grandson. This man, Erich von Tschermak, would go down in the history of science as one of Mendel's staunchest defenders, while his grandfather, a powerful man in his day, would be remembered chiefly as one of Mendel's most ardent adversaries, the man whose defense of the spermist theory of inheritance might have been the spark that fired Mendel's work.

Whatever the explanation, Mendel responded to his examination debacle the way he had to so many other setbacks in his life. He took to his bed. This time, his rescuers were his father and uncle, Anton Mendel and his brother Johann. When they heard about Gregor's second test failure, they rushed to Brünn to try to nurse him back to health. It was Anton's first visit to the monastery, and it would turn out to be his last.

While Mendel was in Vienna to take his exam, he almost certainly paid a call on Professor Unger. We can imagine the two of them, in early May of 1856: the fifty-six-year-old radical botanist, recently reprieved from the threat of being fired for his heretical scientific beliefs, and the thirty-four-year-old monk, desperately trying to qualify as a schoolteacher, the only real job he ever had. There they are, the jaded professor and the teacher with dreams of immortality, sitting in the older man's office, smoking cigars. Maybe Unger offers his guest a glass of tea, or perhaps something stronger. As they speak, Mendel can see the spark of revolution in Unger's eyes. Perhaps he sees, too, that if he is concerned with his own legacy — which seems more uncertain now, the teaching certificate again having slipped from his grasp — he will do well to generate some comparable spark of his own.

Unger's previous few months had been eventful and disturbing. In February he had been on the verge of being dismissed from his teaching post. He had angered a local Catholic journalist by stating that plant species were not fixed but in a state of flux.

"Who can deny," Unger wrote in one of his weekly "Botanical Letters" in *Wiener Zeitung,* the Vienna newspaper, "that new combinations arise out of this permutation of vegetation, always reducible to certain law-combinations, which emancipate themselves from the preceding characteristics of the species and appear as a new species."

Such statements infuriated Sebastian Brunner, editor of the Catholic newspaper *Wiener Kirchenzeitung.* When professors at "so-called Catholic Universities deliver lectures on really beastly theories for years on end — and instruct young people in a nature and world view which is the self same world view that was taught by freemasons with reason everywhere before the French revolution," the mind "boggles," Brunner wrote, following up with a series of dots and dashes to express his unprintable sputterings of outrage.

In February 1856, egged on by Brunner, the university called for Unger's dismissal. Four hundred students rallied to his defense.

The minister of education, Graf Leon Thun, later wrote that the Unger proceedings were the noisiest he had ever attended. In the end, the noise prevailed. Thun allowed Unger to keep his job and forced Brunner to publish a full apology in the Vienna papers. By early March the drama was over.

We can picture Unger telling his former student about his trials, which highlighted an emerging controversy, then seeping through Europe, about which Mendel might have been unaware — the debate over how new species arise. Although Darwin had not yet published his startling *Origin of Species* — that would not happen until 1859, still three years away, and the book would not appear in German translation until a year after that — the idea of evolution was already percolating in intellectual circles. At the University of Munich, for instance, Karl von Nägeli was publishing essays that addressed the question of speciation. "Species cannot exist in complete repose," he wrote. Progressive change "must be perennially at work, and this change cannot fail, in the end, to bring about the disappearance of the species or its transition into another."

Maybe Mendel set out, in his pea experiments, to confirm this idea of "perennial" progressive change as a driving force for the appearance of new species out of the old. If he designed the experiments correctly, he could lend empirical support to the theories of the two botanists he had come most to admire: Nägeli and Unger. Both believed that new species arose out of, as Unger put it, new "permutations" of plant characteristics, subject to "certain law-combinations." Could Mendel prove this in the garden? If he watched and counted for a long enough time over enough generations, might *Pisum* exhibit the "law-combinations" that governed the emergence of new species through the reconfiguration of old, familiar traits?

It may seem, to modern eyes, contradictory for a priest to set about trying to prove the nonfixity of species. A literal reading of the Bible has brought many religious fundamentalists, especially in contemporary America, to a belief that all the species now inhabiting the earth were put here by God during the six days of creation. Even in Mendel's time the Vatican — his employer — was compiling a *Syllabus of Errors* to demonstrate the literal truth of the Bible's Genesis story. Pope Pius IX was conservative in most other ways as well, seeing no need, as he put it, to "reconcile and harmonize himself with progress, liberalism, and recent civilization."

But the Catholic Church in Moravia, as in much of central Europe, took an almost opposite position from the Italian pope's. Central European Catholics were trying as hard as they could to catch up with progress, liberalism, and recent civilization. One of the ways in which the Church was straining to modernize was through the pursuit of the natural sciences. Witness the scientific undertakings of the brethren of the St. Thomas monastery and of other clerics throughout the Hapsburg Empire: they formed the core of a progressive, democratic Catholic intelligentsia in Moravia that managed to ask scientific questions unfettered by Christian dogma. These enlightened clerics, Abbot Napp prominent among them, helped contribute to the unprecedented flourishing of science in Vienna, the hub of the empire, in the second half of the nineteenth century. They created an environment in which it was safe, even laudatory, for a monk like Mendel to take on some of the most freighted scientific concepts of the day.

It may be that Mendel, when he first started working with peas, was not really interested in evolution at all. Maybe he had another score to settle — the one with Eduard Fenzl. If he could demonstrate that both parents were equally involved in passing on inherited character traits, he might at least disprove the philosophy of the preformationists — and might be vindicated in the eyes of Fenzl and his other Viennese examiners.

Or maybe Mendel had entirely different reasons for entering into his investigations — deep-seated reasons, buried in the subconscious. At the time this was still a dark netherworld that no one had the vocabulary to explore. Sigmund Freud, after all, was an infant, born in Moravia the very same week that Mendel was in Vienna quaking through his examination. If we are looking for motivation, Freud might have offered one *post hoc*. Perhaps what drove Mendel was nothing more nor less than the desire to symbolically slay his father and all those father figures — Abbot Napp, Professor Unger, the brilliant and disgraced Klácel — to whom he felt so inferior. He was feeling even more inferior after his second examination failure. Could he get the upper hand by doing at least one thing right, by showing these men, as well as the great indifferent world, that he was an original, creative scientist?

During this darkest of springs, Mendel's peas provided him with his only bit of sunshine. The peas grew in a section of the monastery's large, south-facing kitchen garden, which got sun for most of the day. At the most wilting time, in late afternoon, the garden was sheltered by shadow cast by the brewery, offering a few hours of respite for garden and gardener alike. When the wind blew a certain way, brewery scents mingled with the plants' perfume, peppering the air with a strange mixture of pungency and sweetness: part lilac, part loam.

The garden provided the staples of the refectory table: rows and rows of cabbages, carrots, parsley, pumpkins, and, Mendel's favorite, cucumbers. "Plant more cucumbers — I'm coming home!" he used to write to his family whenever he ventured north for a visit to Heizendorf. He doubtless ate cucumbers just as heartily when they were served, fresh picked or pickled, at the monastery in Brünn.

At the northernmost end of the garden, the new greenhouse sparkled. Sited perpendicular to the library wing, it was long and narrow, with a brick portico at either end lit by huge arched windows and, in between, a vast expanse of glass to trap the sun's heat and keep the greenhouse warm and humid all year round. Inside, in the diffuse light of the south-facing window wall, Mendel arranged dozens of potted plants in neat rows on the floor, on the shelves, on the long wooden tables. The thick air practically tasted green.

Having the new greenhouse and access to the plot in the kitchen garden should have been enough to lift Mendel's drooping spirits. But Napp offered another gift as well. In the autumn of 1855, he turned the rickety glasshouse that Mendel had been using into a handsome orangery — an outbuilding, common then in European gardens, that provided a winter haven for tropical plants. Like most orangeries, it was made of brick and glass and heated by a wood-burning stove to provide warmth and shelter for the monastery's most delicate plants — orange trees, lemon trees, pineapple plants — that could not otherwise survive the icy Moravian winter. In the summertime, when the tropical trees were moved outdoors, Mendel could spread out his work and his thoughts and his pea harvests to his heart's delight. In winter he still could occupy the orangery; he simply had to share it with the potted plants.

The orangery eventually became Mendel's favorite spot in the monastery. He furnished it with a game table for playing chess; an oak writing table; six rush-bottomed nutwood chairs; and a few paintings. In his last years, when as abbot he could use the monastery's grandest rooms, he still spent his time in the orangery — tending plants, playing chess with his nephews, receiving informal guests, and working through the mathematical, biological, and meteorological problems that vexed and intrigued him all the days of his life.

By early June of 1856, one month after the exam in Vienna, and after his father and uncle had returned to Heizendorf, Mendel was his old self again. Indeed, he was better than his old self. After growing peas for two years to be sure he had a parent generation that most decidedly would breed true, Mendel had thrown

Patience to wait out generations

out a dozen or so strains and was left with twenty-two purebreds, all capable of producing, through several generations, offspring that looked exactly like them. Now his hybridizing was proceeding just as he had hoped, his teaching work was satisfying, and his days passed in a swirl of rewarding fellowship and experimentation.

He had also taken on an additional responsibility, one that increased his sense of involvement in the cultivation of scientific knowledge. He was the official weather watcher for the city of Brünn. The job had belonged to Mendel's friend Dr. Olexík, the elderly chief physician at St. Anne's Hospital. But in early 1856 Olexík became too sick to keep up his duties. He asked Mendel, first as an occasional favor and then on a permanent basis, to take local meteorological readings every day and send them monthly to the Vienna Meteorological Institute.

For the next twenty-seven years, almost literally until the day he died, Mendel recorded weather data three times every day: at seven A.M., two P.M., and nine P.M. He would walk over to a corner of the abbot's residential wing and take readings from the thermometer and barometer hanging on the north wall, parallel to the church. After accumulating a month's worth of measurements, carefully recorded in his tiny, perfectly aligned script, he would compile the averages and trends and send all the data to the central weather station in Vienna.

On a cool spring evening in May, Mendel's reading would take place in a garden not yet dark, but bathed in a vibrant blue that would seem somehow lit from behind, even if the waning moon was barely a slit. In February, the month whose name on the Czech calendar, *unor,* means "hibernation," the evening garden would be black and forbidding, and only a deep sense of obligation would drag Mendel to his instruments. In a way, his dedication to this task was fortuitous, because the fame Mendel longed for would come to him in his lifetime, to the extent it came at all, primarily as a local meteorologist.

6

Crossings

The flower sets man a gigantic example of insubordination, courage, perseverance and ingenuity.

— *News of Spring,* Maurice Maeterlinck, 1862–1949

IN A CORNER of the monastery garden, Mendel huddled myopically over rows of greening plants. These were functional little vegetable plants, but they held a strange beauty. Climbing along sticks and strings, they twirled and twisted gracefully as they arched toward the thin Moravian sunshine. The tendrils of these common garden peas, *Pisum sativum,* were at once delicate and tenacious, giving the plants a look of whimsy, though they were among the most utilitarian members of the legume family.

As he moved from one row of pea plants to another, Mendel carefully lifted the leaves away from the slim stalks, pulling at the flowers hidden beneath like coy little butterflies. The leaves were smooth ovals, reminiscent of cupped hands enclosing something precious; as they gradually unfurled, they revealed a flower, either white or a variation of purple, shaped like a tiny bonnet. These flowers, like those of most plants, were the plant's ovaries. Over time the petals faded and the calyx toughened and elongated, becoming the long, leathery pea pods we recognize easily. Within each pod were six or seven peas, the offspring of the plants on

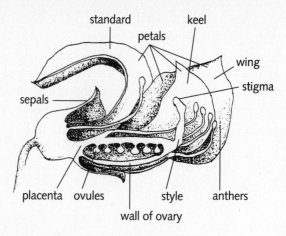

A diagram of the Pisum sativum *flower.*

which they grew. (In *Pisum* the terms "pea" and "seed" are interchangeable.) These peas were the first wave of the next generation.

Almost invariably, the peas looked different from one another, even if they had begun life in the very same flower. This is why the phrase "alike as two peas in a pod," to mean identical, is not exactly accurate. If the expression stayed true to its horticultural lineage, it would refer only to two things that are as alike — or unlike — as any two siblings.

In May of 1856 Mendel was creating a new kind of pea — a hybrid. As he walked up and down the garden rows, he was getting ready to crossbreed two different strains, one of them perfectly round, the other angular — or, as it is commonly and imprecisely translated, wrinkled. This phase of the experiment had begun two months earlier, when Mendel first got ready to sow his round and angular peas. He had taken to heart the old Moravian saying "He who does not cultivate his soil by Gregor's Day is a lazy man." Gregor's Day, which falls in the Catholic calendar on March 12, seems an early date for cultivating, especially in Moravia, where the earth at that season is often still frigid and hard. But that is the folk saying that people abided by.

It seems strangely providential, on reflection, that Gregor, Mendel's religious name, was already enshrined in the gardening lore of his own ancestors. Or maybe there was some element of predestination, of genetic or cultural inevitability, in this man of peasant stock turning his curiosity toward the very questions that had bedeviled his people for generations — the questions of how new and better types of plants are created, how they persist, and how they can be maintained.

Now, in May, in the cool of the early morning — these experiments had to be done before six A.M., when the pollen was still undisturbed — Mendel walked among his pea plants getting ready to crossbreed them. Here he was, a farmer's son whose temperament had driven him from the farm, doing what was at heart an agricultural experiment.

As he set out on this exciting scientific adventure, Mendel would soon encounter surprising truths that in his most fanciful imagining he would not have predicted, truths that would change the face of biology for all time, profoundly affecting how mankind related to its own potential, strength, and destiny.

Mid-spring in Moravia is the perfect time for crossbreeding; the flowers of the pea plant are just emerging but have not begun to bud. As Mendel walked amid the rows of peas, he held a pair of tweezers in one pudgy hand and a camel's hair paintbrush in the other. Bending down with some difficulty, he peeled open the inner layer of the two-part flower — known as the keel — to reveal its male portion, the stamens, made up of long antennalike filaments, each with a tiny yellow bulb, the anther, on top. The anthers contained the powdery yellow pollen, the source of the male gamete. With his tweezers, Mendel pulled the anthers off the flower, thereby removing the pollen and effectively castrating the plant.

How odd for a monk to be desexing pea plants. But Mendel had never shied away from the earthy facts of life. He had, after all, grown up on a farm, where the rhythms of mating and breeding were part of the daily routine. Some of his students, young boys on the prow of puberty, sometimes found Mendel's lectures on reproduction embarrassing and cause for comments or giggles. When this happened, the gentle priest would respond with a rare outburst. "Don't be stupid!" he would shout out curtly. "These are natural things!"

On this morning in May, Mendel slowly moved along his first row of pea plants, emasculating one plant after another, depositing the pollen bulbs in a pocket deep in the folds of his black robe, to be discarded later. Emasculation was necessary because *Pisum*, like a great many other plants, is a hermaphrodite: each flower contains the reproductive apparatus for both sexes. The female pistil comprises the stigma, which catches the pollen; the style, through which the pollen travels; and the ovary, where the ova reside and are fertilized. Under ordinary circumstances, *Pisum* simply pollinates itself. Within the protection of its little covered keel, left to gravity and the passage of time, the pollen from one of the anthers drops down onto the sticky stigma and passes down the style into the inner recesses of the flower, where the female ovary is waiting.

By opening the flower buds to cut out the anthers, Mendel risked disturbing the stigmas while they were still immature. They would need a few more days to ripen to the point of sufficient size and stickiness to retain the grains of pollen that Mendel intended to brush onto them. To keep the buds undisturbed for those few days, he covered them all with little calico caps.

Maybe he was wearing a cap himself, having heard the first notes of the wind chime that hung in the garden. Although he was still a young man — thirty-four on his next birthday, in less than two months — Mendel was in frail health and susceptible to drafts. He had hung an Aeolian harp (named after Aeolus, the Greek god of the wind) to warn him of the sudden gusts that often swept in unexpectedly from the Spielberg hills. Whenever the harp sounded, he would quickly cover his head.

The priest worked slowly and carefully, but his progress was steady. When he reached the end of the first row, he moved on to the next, whose plants were destined to act as the male parent in his breeding scenario. The plants in row two, which Mendel did not castrate, would contribute their male pollen cells to the feminized plants in row one.

A few days after he castrated the plants in the first row, Mendel used his camel's hair brush to sweep the ends of the anthers of the plants that had not been castrated. These would serve as the male parents to the hybrids Mendel was creating. He carefully collected wisps of powdery yellow-orange pollen on the brush's tip. Then he walked over to the feminized plants and dusted the stigmas with his brush with as much delicacy as Georges Seurat would use, thirty years later, to dab the yellow dots suggesting sunshine in his famous Parisian park scene.

After this painstaking brushwork, Mendel placed calico caps atop these plants, just as he had on the castrated plants. This time the caps were to prevent a sudden breeze, or a curious insect, from disturbing the pollen he had deposited so precisely — or from introducing into the mother plant the pollen from a different plant altogether, opening up the possibility that the resulting hybrid would not have the pedigree that Mendel had meticulously created.

Scientists — even those who, like the Moravian monk, seem relatively isolated — build on previous scientists' discoveries and belief systems. Mendel's research was based partly on work done by long line of plant hybridizers. Although farmers had been breeding their crops for at least ten thousand years to create better strains for the market and the table, no one engaged in hybridization for scientific reasons until Josef Kölreuter. This eighteenth-century German botanist made the world's first experimental hybrid plant by crossing two closely related members of the tobacco family, *Nicotiana rustica* and *Nicotiana paniculata*. After two years, he had converted *N. rustica* to *N. paniculata* — a feat that astounded him. "I did not know whether it would be a very much more remarkable thing," he wrote, "if a cat were seen to emerge in the form of a lion."

In Kölreuter's time, the very existence of hybrids conflicted with some of the teachings of philosophy and religion, even if they confirmed centuries of farming experience with plants and animals. "Nature was supposed to preserve the same order and harmony as had reigned in the Garden of Eden," one historian explained. "But if man can create new species whenever he chooses simply by hybridizing existing species there would be no end to the confusion." Kölreuter shared this attitude. So he was relieved to find that almost all of his *Nicotiana* hybrids, despite their lush and apparently complete growth, turned out to be sterile.

What Kölreuter was doing with his experimental plants was little different from what farmers did when breeding fatter cows or hardier strains of wheat. The main difference was in intent. Kölreuter had no plans to bring his plants to the dinner table. What he fed was a person's curiosity, not his belly.

In many ways Kölreuter's quiet, sad life prefigured Mendel's. Like Mendel's family, the Kölreuters were of relatively modest means. Josef's father was the apothecary of Sulz, a little town along the River Neckar in the Black Forest. Like Mendel, the boy was sent, at an early age and at some hardship to his family, to a distant place for schooling. He studied medicine at the ancient University of Tübingen beginning when he was fifteen. Gardening was his passion, though, and he often begged plots of land from friends to conduct his crossings, or grew plants in pots that he could cart around from place to place, a kind of itinerant hybridizer. He abandoned medicine and found no steady work until the age of thirty, when he was hired by the margrave of Baden, Karl Friedrich, and his wife, Caroline, to oversee the royal gardens.

As many men of genius do, including Mendel with his Abbot Napp, Kölreuter had found a true patron. Caroline soon became his most forceful — indeed, his only — champion at court. The head gardener, on the other hand, thought Kölreuter's experiments were not only useless but heretical. Didn't God abhor hybrids? Weren't all hybrid forms chased off Noah's ark? Didn't the Bible state that "thou shalt not sow thy field with mingled seed"?

This gardener, Saul, set about trying to sabotage Kölreuter's plans by ignoring any instructions that were left for him. When Kölreuter tried to grow hybrids in the field, the plot became overrun with weeds; when he tried to grow hybrids in the hothouse, the building turned mysteriously cold. As a result, many of Kölreuter's efforts failed. Still, he managed to conduct 65 successful experiments involving hybridizations among 138 species, which he described in a series of small books published in the 1760s. Although the books remained obscure, much as Mendel's monograph did a hundred years later, they marked the beginning of the systematic, experimental study of plant hybridization.

Saul eventually had the victory he had plotted so venomously. When Caroline died in 1783, Kölreuter, after twenty years with the margrave's court, was promptly fired.

Another German botanist — who, like Kölreuter, came to Mendel's attention in Unger's lectures — had a higher professional status and far more job security. Karl Friedrich von Gärtner (whose name, coincidentally, is German for "gardener") achieved some acclaim in 1837, when he won an essay contest sponsored by the Dutch Academy of Sciences. In that era scientific societies sponsored essay competitions to stimulate thinking along certain lines of inquiry. The process resembled the contemporary American system of Requests for Applications, or RFAs, which are disseminated to the research community to encourage scientists to address particular questions by offering sizable federal grants. In the eighteenth and nineteenth centuries, the triple lure of prize money, publication of the winning essay, and no small amount of fame was dangled, sometimes for years, to help move scientific progress in the direction the sponsoring agency desired.

In this case the agency wanted botanists to think about hybrids — and especially their commercial possibilities. "What does experience teach," the essay question asked, "regarding the production of new species and varieties, through the artificial fertilization of flowers of the one with the pollen of the other, and what economic and ornamental plants can be produced and multiplied in this way?"

Gärtner did his hybridizing in the remote Black Forest village of Calw, squeezing in his gardening work whenever he could steal time away from his medical practice. He did not even hear of the Dutch Academy contest until six years after the deadline. But no one had yet submitted a single entry, so the academy encouraged Gärtner to write up his experience with hybrids. In 1838 he was awarded first prize; his was the only submission.

For the next eleven years Gärtner worked on final revisions of the prize-winning essay, which ultimately described his experience with nearly 10,000 hybridization experiments on a total of 700 species, yielding some 250 hybrid plants. He published the final product, at his own expense, as a massive book entitled *Experiments and Observations Concerning Hybridization in the Plant Kingdom*. Darwin, Gärtner's much younger contemporary, said the book contained "more valuable matter than all other writers put together, and would do great service if better known." But Gärtner, who was already seventy-seven, never saw his reputation grow. The book was his only legacy; he died in 1850, soon after it appeared.

Mendel, though aware of the brilliance of his predecessor, was bold enough in designing his own experiments to recognize the flaws in Gärtner's. He showed the sensibility of a modern scientist, whose *métier* is replication of prior results, when he wrote: "It is very regrettable that this worthy man did not publish a detailed description of his individual experiments, and that he did not diagnose his hybrid types sufficiently." This imprecision meant that he could not reproduce Gärtner's results even "in a single case!" In addition, Gärtner seemed to have had no idea whether the plants he used as the parental strains were pure-breeding or were themselves hybrids. Nor did he carry his crosses into the second and third generations — an essential step if results of further hybrid-hybrid crosses were to be detected.

Also, Gärtner had focused on the wrong unit of experimentation. Mendel always considered his object of study to be the plant's individual parts, its character traits. Gärtner, like so many of his contemporaries, was interested in the plant as a whole, which he considered the expression of all its parts. The thought of examining each part individually did not fit into his overall view. The holistic conception of inheritance, common even into the late nineteenth century, led to the widespread acceptance of "blending inheritance," the theory that offspring were a combination of traits and thus roughly midway between the two parents.

But Mendel found seven different traits in peas, easy to identify with the naked eye, that occurred in an either-or configuration. They never blended; they were always inherited separately and intact. "Transitional forms were not observed in any experiment," he said emphatically. In the "height" character, for example, a plant was either very tall — some six feet or more and needing to be staked — or very short, no more than twenty inches high. No traits — at least of the seven Mendel chose — were in between or blended. He called the traits "pairs of differentiating characters." If he could identify these differentiating characters in his peas and keep track of how they were transmitted from parent to offspring, then, he hoped, he could "deduce the law according to which they appear in successive generations."

Mendel's seven "character traits" were as follows:

seed shape: either angular ("wrinkled") or round
seed color: either green or a variation of yellow
tint of the translucent seed coat: either white or gray
shape of the ripe pod: either inflated enough to appear smooth (with the peas inside invisible) or constricted and revealing the bumpy contours of each pea
color of the unripe pod: either green or yellow
location of the flowers: either restricted to the tip (terminal) or distributed evenly along the whole stem (axial)
height: either tall or dwarf

Did Mendel distinguish, in his own mind, between the traits you could see — a plant's height or pea color or type of pod — and the units that created those traits? It seems a central question, since the idea of the units that underlie character traits, now known as genes, has become a basic concept in the study of inheritance. But no one knows for sure what Mendel thought about underlying units — or even whether he thought of them as discrete units at all. Insightful as he was, he was still a product of his times, and the prevailing belief about heredity was that if there were units, they were not necessarily solid particles. Mendel might have seen his underlying traits as amorphous, shapeless things, sort of like blobs of molasses, that took on the form of whatever vessel they happened to be in. Or maybe he saw them as units that were free-standing and intact. However, no matter what analogy he used to envision his units, it no doubt was far removed from the contemporary conception of the gene.

Mendel used two different words that are both usually translated as "character trait." But in German these words have slightly different shades of meaning. The first, *Merkmal,* implies a quality you can see and recognize, something we usually call a "trait." The second, *Elemente* — which Mendel used only in the plural form — is close to its English cognate, "elements," the unknown substances that might account for an organism's *Merkmale.* Mendel used the word *Merkmal* more than 150 times, compared to just 10 for *Elemente,* which he used only in the conclusion of his paper, where he deduced units, or elements, from the way the traits he had been observing were passed on from generation to generation.

The men who translated his paper into English in the early 1900s might have been overly generous in their assumptions of what Mendel understood. In their translations *Merkmal* often became "unit" or "factor" or "determinant." With twentieth-century hindsight, the translators' readiness to ascribe to Mendel all sorts of prescient views of heredity through the use of these modern words may have made it seem that Mendel was closer to the concept of a genelike particle than he actually was.

It turned out that the seven traits Mendel noted were a particularly lucky set, since they were always transmitted independently of one another. Unlike many other traits — such as flower color and pollen shape in the sweet pea, which tend to be inherited together — the height of a *Pisum* plant said nothing about, say, its seed color: tall plants were as likely to bear green peas as they were to bear yellow and were as likely to bear yellow peas as dwarf plants were. The reason for this nice clean pattern was not revealed until nearly a century after Mendel's death, when a map of *Pisum's* seven chromosomes was finally pieced together. Each of Mendel's seven traits appeared on a different chromosome — or, in one case, on two distant ends of the same chromosome. This reduced the chances that these traits would be coupled through a process called linkage — which would have seriously muddied the monk's results.

Of course, the chromosomal distribution of these seven traits might not have been a matter of luck at all; Mendel might have made his own luck through careful, methodical planning. As he later described it, his earliest crosses involved not seven pairs of character traits but fifteen. His reason for ultimately discarding the results of half of the experiments might have been because those traits did indeed link with each other, leading to muddy and patternless results that could not be explained.

Mendel began by studying seed shape. He knew, from his two years of work with true-breeding strains, that the seeds were always either round or angular. No doubt he anticipated, as he crossed his first peas on this May morning, exactly how the first seeds would look when he collected them a short time later. He had already read and reread Gärtner's book, a dog-eared copy of which was in the monastery library, darkened with marginal notes in Mendel's distinctive hand. Like other hybridizers, Gärtner had shown that all members of the first generation of hybrid plants look alike, resembling either one parent or the other. The trait revealed in this way would come to be known as the "dominating" trait, a term Mendel introduced in 1865. Round was the more common seed shape in *Pisum*, and Mendel predicted that the first generation of hybrid peas all would be round.

But, skilled chess player that he was, he was thinking through his next move and his next, even during this first spring of crossbreeding. He wanted to see what would happen to the non-dominating trait, the one that in the first hybrid generation seemed to disappear, which he would later call "recessive." By the following autumn, the peas in the garden would be the offspring of the plants he crossbred now. They would be his first hybrids. This was the generation that subsequently came to be known as the F1 (for "first filial") generation. Mendel himself never used the term, but his twentieth-century advocates did, and it makes clear the relationship among the first hybrids and their descendants, the F2's, F3's, F4's, and so on.

The schedule of planting, fertilizing, harvesting, and counting fell into a seasonal routine: by Gregor's Day of 1857, cultivate the soil; in early spring, plant the round F1 seeds; let them self-fertilize in the late spring; that summer, harvest the pods, still part of the F1 generation, and split them open to reveal the peas inside, the earliest dispatches from the F2. Begin the cycle again on Gregor's Day of 1858 with the F2 peas, and on Gregor's Day of 1859 with the F3's.

The planting and harvesting schedule, though it had so much waiting built in, was still relatively speedy because, in these first experiments, Mendel was looking specifically at the characteristics of the peas themselves — the offspring of the plants on which they grew. Had he been looking at the traits of the flower or the plant or the pod, as he would be doing in subsequent years, his wait would have been measured in months rather than weeks. To record characteristics of next year's *Pisum* plants or flowers, you first have to wait until this generation's peas have dried out and the earth has become sufficiently soft to sow the seeds in early spring. That means a delay of at least nine months after the summer harvest until the next generation can even begin. Then you have to wait until the plants grow and the flowers bloom, which takes another month, possibly two.

According to this plan, Mendel's first meaningful results would be in hand by the autumn of 1857. He would then know what kinds of offspring his round/angular hybrids had created. And, if his hunch was correct, he would be on his way to developing a law of inheritance that worked not only for peas but for beans and snapdragons, mice and maple trees, lizards, wolves, and people.

7

First Harvest

What is a garden for? For . . . visions of the invisible, for grasping the intangible, for hearing the inaudible.

— *Our Gardens*, Samuel Reynolds Hole, 1819–1904

MORAVIAN PEA PLANTS produce fat, slightly waxy pods that fit snugly in the palm of your hand. At a little over three inches long, each one nestles there as if specifically designed to do so. When you bring the pod indoors, away from the sun, it becomes cool to the touch. A little jester's cap at one end marks where it was pulled off the stalk, and the peas inside feel like marbles in a leather pouch. To split the pod open — a sharp fingernail helps get you started — you need two hands: one to cradle it, the other to do the splitting. Open, it bursts with the scent of grass. Each pod holds half a dozen peas, sometimes more, and each makes a tiny, nearly inaudible "pop" when you pull it free.

This cradling, splitting, and popping would occupy Mendel over long evenings for a long, long time. During his harvest seasons — autumn if he was looking at the traits of peas or pods, early summer if he was interested in characteristics of flowers or the whole plant — he often would take his samples, sorted into bags labeled with the row and plant from which each one had come, to the orangery. There he could sit in the calming warmth of the little front room, easing himself into a cane-seated chair pulled up to the ta-

ble. Each fall between 1856 and about 1863, his work involved freeing and categorizing the peas: grab a pod, hold it in one hand, slit it
open with the other, pop out the peas, classify each one, keep a tally
of which pod and plant it had come from, distribute the peas to
their properly labeled bags, grab another pod, hold it in one hand,
slit it open. And so on and so on.

No question about it: this was tedious work. In the fall of 1857
alone, when the first F2 hybrid seeds were finally in hand, Mendel
had to shell, count, and sort by shape more than 7,000 peas. And
that was just for one experiment, involving crosses between round
and angular peas. By the time he was done, seven years after he
began, Mendel had conducted seven versions of this experiment,
seven different "monohybrid" crosses, designed to look at plants
that varied according to only a single trait (shape first, then color,
then height). He also conducted two "dihybrid" crosses, in which
he crossed plants that varied according to two traits, to see which
hybrids arose and in what proportions. In a dihybrid cross, he
crossbred plants bearing, for example, yellow, round peas (double
dominants) with plants whose peas were green and angular (double recessives). Later, he did a "trihybrid" cross, the most difficult
to analyze, in which he crossed a triple-dominant plant (yellow,
round peas with gray-brown seed coats) with a triple-recessive
(green, angular peas whose seed coats were white). Both the dihybrid and trihybrid crosses answered a particular question: do traits
pass from parent to offspring in tandem or separately?

After doing all these hybrid crosses, Mendel tried an experiment
that was the ultimate test of his emerging theory: the backcross. He
held off on doing this until the last year of his experimentation,
because it was with the backcross that his ideas about inheritance
at last would be confirmed — or refuted. By the time Mendel was
done with this succession of crosses, recrosses, and backcrosses, he
must have counted a total of more than 10,000 plants, 40,000 blossoms, and a staggering 300,000 peas.

In early autumn of 1857, Mendel glimpsed the first peas of the F2 generation. He knew the breakdown between round and angular peas could be the first visible proof of his ideas about plant hybridization and species formation. He had to work to keep himself from taking shortcuts in his eagerness to see if he had been right.

What was Mendel really thinking that autumn? What was he expecting to find in those F2 peas? Science textbooks tend to present Mendel as a pioneer in the deductive scientific method: growing, harvesting, and counting his peas to test a clearly formulated hypothesis regarding the mechanisms of inheritance, especially the random recombination of dominant and recessive particles. But despite the legends that have built up about what he was doing and thinking in his garden — legends created in part by his rediscoverers and in part by Mendel himself — what took place was probably much muddier than that.

Mendel probably believed that counting his pea progeny would reflect an underlying mathematical relationship. He got this idea from his math and physics training at the University of Vienna, as well as from attitudes circulating at the time. This was, after all, the middle of the nineteenth century's "avalanche of numbers," and a hallmark of Mendel's genius was his receptivity to the moods and methods of his generation. So, like so many of his contemporaries, he counted. But Mendel counted more than most other botanists did. He applied his passion for counting, which helped him reach his breakthrough conclusions about inheritance, almost indiscriminately to everything in his own little world. He counted not only peas but weather readings, students in his classes, bottles of wine purchased for the monastery cellar. He counted because that was what was done in the mid-nineteenth century, and because he had an abiding faith in the clarity of numbers.

This does not mean he knew what would result from counting

his F2 hybrid peas — nor that he knew how to explain the mathematical expressions he ultimately detected. Indeed, once he was
able to make numerical sense of his results, he almost certainly
could not offer, at least at first, a full explanation.

Eagerly, Mendel shelled the peas from the F2 generation, the offspring of the 250 F1 hybrids he had allowed to self-fertilize the
previous spring. He counted 7,324 peas in all. Most of them (specifically, 5,474) looked just like the peas of the F1 parent hybrids.
They were round. But the rest were angular, like some of their
grandparents. This angular group comprised 25 percent of the F2
generation: 1,850 peas.

The appearance of those 1,850 angular peas in the F2 generation
showed that the angular trait had not disappeared in the F1's; it
had only been hidden. Mendel did not know yet how or why, but
he was on his way to finding out. What he did notice, and tuck
away in his mind, was that a whole-number ratio could describe
the proportion of round peas to angular peas among the F2's. The
ratio was 3:1.

Next Mendel studied a second trait, pea color. To do so, he probably resorted the same seven-thousand-plus peas by color instead
of shape. This shortcut saved him a year of waiting, but it involved
a lot of forethought. Mendel could not have resorted his peas without having developed an elaborate system of labeling, beginning in
1856, to keep track not only of the parents' shapes but of their colors and possibly other traits as well. If Mendel did in fact reuse his
F2 peas, for a second and maybe even a third or fourth experiment,
we can only assume that he had planned thoroughly enough to
keep meticulous records from the very start.

In the second experiment, the F1 hybrid generation all bore yellow peas. In the F2 generation, which Mendel probably analyzed in
the fall of 1857, most of the peas (6,022 of them) were yellow. But,

as Mendel expected, a subgroup of green peas (numbering 2,001) also appeared. As happened with pea shape, the trait that had seemed to disappear among the F1's (in this case, green color) turned up again in at least some of the F2's.

Once again almost exactly one pea of every four revealed the previously hidden trait. Once again the proportion of F2 peas with the visible trait in relation to those with the hidden trait could be expressed as a ratio. And once again the ratio was three to one. The ratio "relates without exception to all the characters which were investigated in the experiments," Mendel said.

But what did this 3:1 ratio mean? Where did it come from, and what did it say about the way these traits were inherited? Answering these questions became Mendel's primary goal. He could not really refine a theory until he had allowed his hybrids to self-fertilize one more time, leading to a new generation, the F3's, ready to be counted in the fall of 1858. But even before then he could make some conjectures.

For one thing, he could differentiate between the two traits by calling one "dominating" and the other "recessive." The dominating trait — which we now call "dominant" — was the one that showed up in all the F1 hybrids and three-quarters of the F2's. The recessive trait was the one that seemed to disappear entirely in the first filial generation, but arose again in one-quarter of the second.

Considering one trait at a time, Mendel represented each with a letter of the alphabet. The trait could be passed on in one of two forms — in the case of pea color, for example, either in a form that would lead to yellow or one that would lead to green. Mendel used a single letter to represent each characteristic, with a capital letter (A) to indicate dominance, a lowercase letter (a) to indicate recessiveness, and one of each (Aa) to indicate the hybrid.

Using binomial letters was a brilliant stroke. No one else had ever used anything like this system. But it is not clear what exactly Mendel meant the letters to convey. While other botanists, including Nägeli in 1865, used letters to represent certain characteristics, Mendel was the first to use double letters for his hybrids — indicating that he understood that a hybrid carried two differentiating character traits, one of which was hidden from view until it chanced to reappear in subsequent generations in a different combination. This made it seem as though Mendel knew that each individual would possess two of these traits, or *Merkmale,* one from each parent. Yet in the pure-breeding plants, either pure yellow or pure green, he used only a single letter to represent the full complement. A pure-breeding yellow was represented with a capital *A,* a pure-breeding green with a lowercase *a.* Did Mendel believe that only one such unit existed in these purebreds? Or, as is more likely, did he fail to see his letters as representing any sort of particle-like unit at all, but something more formless, like a dollop of honey added to the doughy mixture of an organism? In that case, he would have seen no need to double the letters if both were the same, since two bits of formless matter would behave in almost exactly the same way as one bit would.

Mendel's contemporaries would not have been able to help him much in perfecting his binomial code. The man who might have come closest was Charles Naudin, a French botanist from the Museum of Natural History in Paris. But while Naudin nearly approached a Mendelian understanding of hybridization, his theories were quite different from the monk's. He believed hybrids contained two "essences" in each of their sex cells. If he had used letters, he probably would have said that each of a hybrid's gametes could be represented by the two letters *Aa.* Naudin's view of hybrids did not include separation in the sex cells.

Naudin's work won international recognition in 1862, when he won an essay contest sponsored by the Paris Academy of Sciences, which was looking for a description of plant hybrids "from the point of view of their fecundity, and of the perpetuity or non-perpetuity of their characters." But his scholarly success came at a time when his career was crumbling because of a nerve disorder that left him totally deaf and constantly in pain. Forced to resign his post as professor of zoology at the Collège Chaptal in Paris, he spent the rest of his long life (he was eighty-four when he died in 1899) carving out a living by selling garden specimens and seeds.

In 1860, when the French competition was announced, Naudin had been studying hybrids for years. Much as he loved them as experimental subjects, he thought hybrids were deviant and unnatural. He saw Mother Nature as an antimiscegenationist intent on destroying mixed breeds whenever they appeared. "Nature is eager to dissolve hybrid forms which do not enter into her plan," he said, as made clear "by the separation of the two specific essences [in each hybrid germ cell] which art or chance has violently brought together." The tyranny of Mother Nature, or God, was explanation enough for Naudin of one of his most puzzling observations: that the offspring of a hybrid *Primula* (primrose) completely reverted to one of its parental strains. He saw it as nature trying to get back to its original purity. Mendel, on the other hand, looked to mathematics rather than God to help make sense of such things; to Mendel, reversion to an earlier form was simply the reappearance of a recessive character trait.

Despite his blind spot, Naudin had some brilliant insights into the behavior of his hybrid *Primulas,* and he nearly arrived at the law of segregation three years before Mendel did. But his unwillingness to count and calculate his primroses — making him a scientist in the nineteenth-century mold of collector, observer, theorist, operating along the same lines as Charles Darwin — finally cost him dearly.

During his lifetime, the very fact that Naudin was so much a product of his age seemed a blessing. His contemporaries respected him as a scientist of great insight and originality. Darwin himself read and admired his prize-winning essay, and entered into a correspondence with Naudin that lasted from 1862 until Darwin's death twenty years later.

We can see today that Naudin, the famous one, was far less lucky than Mendel, the obscure. Naudin's mathematical blind spot, though it made him more accessible to his contemporaries, rendered him little more than a footnote in the history of genetics. He lacked Mendel's insight that the hereditary factor (the *Merkmal*) was a separate and distinct unit passed on individually. If he had been able to go the extra step and bring math and statistics into his analysis, we might today be studying Naudinian rather than Mendelian genetics.

But Mendel knew better. He knew, and twentieth-century genetic research has borne him out, that each parent passes on only a single *Merkmal:* either *A* or *a*, but never both. So the quiet monk, it turned out, had more good fortune in the long run than did his more fêted contemporary.

One of Mendel's most revolutionary insights was that the combinations of *Merkmale* were essentially random. Four kinds of germ cells were available for mixing — *A* pollen, *a* pollen, *A* egg, and *a* egg — and each had an equal chance of joining with each of the others. The *A* pollen was just as likely to pair with the *A* egg as with the *a* egg; similarly, the *a* pollen had an equal chance of pairing with an egg from either *A* or *a*.

Mendel's calculations allowed him to go one step beyond his simple 3:1 ratio. When he thought through how the next year's peas would look, he understood that among the F2 peas arrayed on his table in the orangery, there were really two types of yellows. Of all the yellow peas (which made up three-quarters of his entire cache), one-third would "breed true" in the next generation, producing only yellow seeds. Two-thirds would behave just as their parents had and produce two different colors, yellow and green, in the 3:1 ratio again.

It is hard to believe that Mendel could have envisioned this notion clearly in the autumn of 1857. It is easier to imagine that he reconstructed the idea in hindsight, after he saw how the crops of 1858 and 1859 and maybe even 1860 actually looked. But the explanation he finally arrived at, long before he formalized his thoughts in the two-part lecture delivered in early 1865, was this: that for every four seeds resulting from self-fertilization of a yellow-green hybrid, one of the yellow peas could be designated as *A,* or pure yellow; two were yellow but also carried the potential to produce green peas, which could be designated as *Aa;* the last pea, the 1 in the 3:1 ratio, would be green, or *a.*

In this way Mendel made the first tentative step toward a concept that would not be fully elucidated for another fifty years: the difference between phenotype (the way something looks) and genotype (the particular combination of genes that explains those looks). If a pea's phenotype is one of Mendel's dominating alternatives — round-seeded rather than angular, or yellow-seeded rather than green, or smooth-podded, or tall — you could not tell just by looking what its genotype was. (This is not true for peas with recessive phenotypes, which are always pure recessives, since dominant traits cannot hide in the genotype but are always revealed in the way a plant or animal looks.) In other words, you could not tell just by looking whether a yellow pea carried only the yellow *Merkmal* for pea color (*A*) or whether it also carried a hidden *Merkmal* that carried the potential for the recessive trait as well (*Aa*). That yellow pea could be either a pure dominant or a hybrid, and only further breeding would tell you which it was.

Mendel eventually carried out his crosses in this particular set of experiments for four more generations, so he had a total of six generations derived by self-fertilizing hybrids. In each generation the seed traits broke down into three types: pure dominants (meaning their descendants were all dominant themselves), pure recessives (meaning their descendants were all recessive), and a

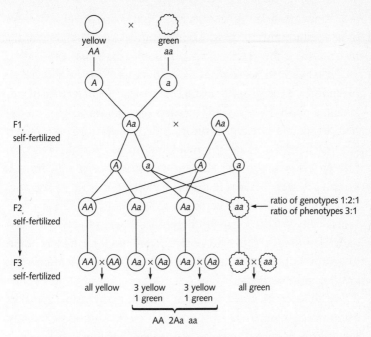

A monohybrid cross between yellow and green peas, using more modern terminology for purebred peas (AA and aa) than Mendel used.

group that revealed themselves to be hybrids only after giving rise to both kinds of offspring, always in that same 3:1 ratio. With the new information derived from successive generations, Mendel could refine the 3:1 ratio he first saw in the F1's. It was really a ratio of 1:2:1 — one pure dominant for every two hybrids for every one pure recessive. In Mendel's shorthand, *A:2Aa:a*.

During the autumn of 1857 Mendel could be seen mumbling to himself as he wandered through his flower bed; humming during his walk to the Brünn Technical Institute, the Realschule, where he taught at least eighteen lessons a week and sometimes as many as twenty-seven; eating heartily, as attested by his expanding waistline; frequently smiling his shy, close-lipped smile. He was swept up in a community of good feeling, enjoying the company of his students, his fellow teachers, his brethren in the monastery.

"I still seem to see him as he walked back to the monastery through the Bäckergasse," recalled one Brünn acquaintance of this golden time in Mendel's life: "a man of medium height, broad-shouldered, and already a little corpulent, with a big head and a high forehead, his blue eyes twinkling in the friendliest fashion through his gold-rimmed glasses. Almost always he was dressed, not in a priest's robes, but in the plain clothes proper for a member of the Augustinian order acting as schoolmaster — tall hat; frock coat, usually rather too big for him; short trousers tucked into top-boots." His dress bespoke his decorum and modesty; he was out in the world, but always as a cleric who had taken vows of chastity, poverty, and reverence.

"We all loved Mendel," a student recalled many years later, after his former teacher had been heralded as a hero of modern biology. "I recall his dear, loyal face," said another, "his kindly eyes which often had a roguish twinkle, his fair, curly head, his rather squat figure, his upright gait, the way he always looked straight in front of him; and I hear the sound of his clear voice, note his strong Silesian accent." Sometimes these same schoolboys severely tested their teacher's famously mild manner. According to one story, Mendel stuffed his pockets each morning with peas, so he could throw a handful at any poor student who dared to nod off in class.

If his classes were held in late morning and he kept his windows open, Mendel might have been able to hear the pealing of the noontime bells of the Cathedral of Saints Peter and Paul at the in-accurate hour of eleven A.M. The odd timing carried, as so many traditions do, a story. The year was 1645, the setting the Thirty Years War. General Linart Torstenson and his Swedish army had been occupying Brünn for weeks but had not been able to overrun the town as they had so many towns and villages throughout

Moravia. In Brünn the Swedes and the locals were locked in endless stalemate.

One morning, in frustration, General Torstenson declared that if his troops had not won by midday, they would finally retreat. This seemed to give his army renewed fighting vigor, and by midmorning Brünn was on the verge of collapse. The bell ringer, hoping either to rally the townspeople or to outwit the general — his motivation is now one of history's misty uncertainties — climbed the clock tower at eleven that morning. One hour too soon, he rang the midday bells, and General Torstenson, true to his word, gave up the battle and left. This moment of victory has been relived every day in Brünn right up to the present, more than 350 years after the Swedish defeat. This is a town that clearly values its long life story.

The shadows grew shorter in the garden, and Gregor Mendel straightened and removed the hat he had slapped on to protect him from the morning breeze. Midday was approaching, and the priest was hungry for dinner. As he collected his brushes and tweezers, perhaps he was humming to himself and muttering, as he so often did after a few hours with his pea plants. Working on *Pisum* always seemed, in a strange and comforting way, to settle him.

No matter what else happened in his secular or religious life, Mendel always found peace in his garden. A day spent in the garden, he said, was a kind of resurrection: "Every day from spring to fall, one's interest is refreshed daily, and the care which must be given to one's wards is thus amply repaid."

The extent to which it would be repaid was yet to be uncovered.

8

Eve's Homunculus

From yon blue heavens above us bent,
The gardener Adam and his wife
Smile at the claims of long descent.

— *"Lady Clara Vere de Vere,"*
Alfred, Lord Tennyson, 1809–1892

TAKE APART A *matrushka*, a Russian nesting doll, and you see a marvel of woodworking talent and near-infinite patience. Not only do the halves fit together perfectly, top to bottom, but each doll is smaller than the previous to just the right degree. The miniaturization itself is ingenious, gradually reducing the scale of the next doll and the next, until the smallest one of all, the one so tiny it cannot be split in two, is almost unbelievably minute. How can a human hand accomplish such detail, paint facial features so tiny, carve a piece of wood that you can barely hold, let alone whittle?

Now imagine that Russian doll on a scale so minuscule as to be microscopic. Imagine the progressive miniaturization going on not just through ten dolls, but through tens, even hundreds, of thousands. Then you are close to imagining how some nineteenth-century biologists explained the enigma of generation.

These "preformationists" pictured a series of nesting creatures analogous to Russian dolls in the gametes of every plant or animal that God had placed in Eden. Each of these creatures was an intact organism — just vanishingly small — with a tiny version of a heart, a brain, a wing, a petal, whatever the full-grown living thing would need. And inside the eggs or sperm of this biological *matrushka* was another series of nesting creatures just like it, minus one, to carry on the process for all possible future generations.

It is like counting the angels dancing, trying to visualize how many preformed beings could be contained in this way and how many it would take to account for every newly born or hatched or blossomed organism in existence, from Creation on. In an apple tree's fruit lay all the apple trees of the future; in a calf's ovaries, the preformed beings of every other cow. And in the ovaries of a baby girl or the testes of a baby boy, hundreds of thousands of generations of one human lived inside another — in a special kind of preformed entity called a homunculus.

In the Bible story of Genesis, God created Adam and Eve, the first humans, on Day Six. First came Adam, created in God's own image, formed of "the dust of the ground"; God blew into Adam's nostrils "the breath of life, and man became a living soul." Then, to be Adam's mate and companion, God took one of Adam's ribs and fashioned Eve. So far as we know, Eve came with all the generative equipment possessed by modern woman — a womb designed to nurture the growth of embryos, a birth canal through which the fully grown fetus could pass into the world, a pair of breasts with milk ducts for nursing the new baby to health and strength.

And inside each of Eve's two ovaries was a homunculus.

What makes us human? Great thinkers have wondered about this since antiquity. Why do we resemble our parents? Why do dogs always give birth to dogs, larks to larks, roses to roses? Aristotle was among the first to answer these questions with a modicum of accuracy, propounding a theory of "essentialism" in the third century B.C. Essentialism held that every part of every new organism is found in the menstrual blood of its mother and activated by the semen of its father. Eventually this idea turned into two schools of thought, both subsets of preformationism: the "spermists," who, like Aristotle, pictured the tiny preformed being inside every drop of sperm; and the "ovists," who put the homunculus into the egg.

Spermists got a boost in 1677, when Anton von Leeuwenhoek, a Dutch merchant and amateur inventor, created the world's first microscope. One of the first things Leeuwenhoek looked at under the microscope was fresh semen, which seemed animated by the moving shapes he called "animalcules." The idea became fixed that semen contained all future generations in its nice, neat, easily transferred little package, the sperm.

With Leeuwenhoek's apparent confirmation of the theory, preformationism retained an even firmer grip on scientific thought for the next century. Its first real challenge came in 1745 in a book called *Venus Physique*. Written by Pierre Louis Moreau de Maupertuis, a leading Newtonian scholar in France, the book suggested that all living organisms had been slapped together by "blind destiny." No wise or benevolent Creator was included in his view of life.

Chance created a huge number of individuals, Maupertuis wrote, but only a handful could survive, those that "were organized in such a manner that the animals' organs could satisfy their needs." These lucky few became the species that ultimately inhabited the earth, representing "a small part of all those that a blind destiny has produced." It was a little like the idea that if you let enough monkeys pound away on enough typewriters, one of them will, by random chance, type out a Shakespearean sonnet.

In Maupertuis's view, male and female contribute equally to the traits of the next generation, which are passed on through both sexes' "fluid semen." Male and female semen combine in the mother's womb, where the mixture may develop into an embryo. Each structure of the embryo forms when the correct portion of the parents' conjoined semen is drawn to its correct spot by Newtonian attractive forces.

In *Venus Physique*, Maupertuis anticipated by more than a century some of Mendel's central ideas: the equal contribution of both parents to the offspring; the random nature of these combinations; the transmission of particular odd characteristics (he looked specifically at polydactyly, the inheritance of an extra finger) from one generation to the next. Just a few years later another French nobleman published a book that was just as prescient, this time anticipating the central ideas of Charles Darwin. The book, *Histoire Naturelle,* was written by Georges Louis Leclerc, comte de Buffon — generally known simply as Buffon.

Buffon was an eccentric, especially during his most creative spurts. In order to write, he had to dress up in his finest regalia, from braided wig to silk waistcoat to a lacy, high-collared shirt. Like Maupertuis — indeed, like most biologists of his time, until they were set straight in the 1880s by another Frenchman, Louis Pasteur — Buffon believed in spontaneous generation. His experiments on meat gravy, which he boiled and sealed in a flask, made him believe that the tiny creatures he saw swarming there a few days later had arisen, as if by magic, from dead organic matter.

Spontaneous generation was, for centuries, a widely held belief about how life began. Hamlet, for example, told Polonius that "If the sun breed maggots in a dead dog," then Polonius should be careful with his daughter Ophelia. "Let her not walk i' th' sun," the half-mad Hamlet advised. "Conception is a blessing, but as your daughter may conceive, friend, look to't."

While Maupertuis believed that Newtonian forces imposed structure on the embryo, Buffon thought the embryo's organization could be traced to an "internal mold," or template. This mold — something not so different from our contemporary understanding of the DNA blueprint — preserved the species from one generation to the next. But a species could also change, according to Buffon's way of thinking. In 1749, when the first of his three-volume *Histoire Naturelle* was published, he wrote that an ancestral form of a particular organism might diverge into a number of species, and that migration to different parts of the world might cause the divergence. And, like Darwin more than a hundred years later, Buffon wrote that the environment acted directly on the organism through organic particles — something quite different from our twenty-first-century view.

Any second-grader knows that if you mix equal parts of yellow paint with blue, you will get enough green to paint all the grass you want. For a long time scientists thought inheritance worked like finger paints: take a blue mother and a yellow father, and you get offspring that resemble neither parent but are a unique intermediate blend: green.

This notion of blending inheritance, that offspring take an intermediate form midway between their parents, prevailed until Mendel's day. Charles Darwin, for instance, believed in it, though the idea caused problems for his theory of natural selection. The theory states that when inheritable changes occur in an organism — and Darwin never did quite figure out how, or even how often, those changes happened — the ones that offered some selective advantage would persist, and those variants would become more and more common as the generations passed. If inheritance were a matter of blending, however, every variant would effectively become blended out, or "swamped," in just a generation or two. Picture it as a drop of red finger paint — representing one of Darwin's small adaptive changes — in a gallon bucket of white. Within a few generations of blending, accomplished by stirring the white paint a few times, the whole bucket looks white again, as though the red paint had never existed.

If Darwin had known about Mendel's work, he would have had a rebuttal to the swamping argument. Recessive traits do not get blended away in Mendel's scheme; they disappear in hybrids that also carry the dominating trait, but reemerge in later generations when the gametes rearrange themselves and form a few double-recessive offspring. Mendel's experiments, which revealed the random and independent transmittal of different traits, threw notions of blending inheritance into question. Character traits cannot be inherited separately — a condition that would later become known as "segregation" — and at the same time also blend.

Some have thought it a great loss to nineteenth-century biology that Darwin did not know about Mendel. They say his arguments about "descent with modification" — Darwin himself never used the word "evolution" — would have been more quickly accepted if he had been able to back them up with a coherent theory of inheritance. Had he read about Mendel's work with peas, Darwin might have devised a mechanism for natural selection with more empirical support behind it than the erroneous hypothesis he eventually put forth.

At first, however, the furor over the appearance in 1859 of Darwin's revolutionary book, *On the Origin of Species by Means of Natural Selection, or the Preservation of Favoured Species in the Struggle for Life,* had little to do with mechanisms. People did not chafe against questions about *how* natural selection occurred; what inflamed them was the very *idea* of it occurring. Darwin's theory cut to the most basic assumptions about mankind's place in the universe and about the hand of God in Creation. Most of his countrymen, as well as people throughout Europe and the United States, had no intention of allowing their most dearly held beliefs to be so badly shaken.

9

The Flowering of Darwinism

In the possibility of change lies the imperishable charm of gardens. Forever through past experiences shine the bright alluring pictures of the future.

— *The Garden Month by Month,*
Mabel Cabot Sedgwick

THE CROWD IN THE LIBRARY was growing ugly. It had assembled, some seven hundred strong, after having been herded first into the lecture hall and then, when that started overflowing, into the long West Room. Even that was not big enough to hold all the people, so now they were in the library. On hand were scientists, theologians, Oxford dons and undergraduates, and even women, those "glorious maidens and matrons" whom at least one scientist, the esteemed Adam Sedgwick of Cambridge, wanted to shield from the horrors of evolutionary thought. The clergy assembled in the middle of the room, the undergraduates in the northwest corner. Except for the students, most of the people in the hall were decidedly anti-Darwin.

It was June 30, 1860, in the middle of the weeklong annual conference of the British Association for the Advancement of Science. This year the group was meeting in Oxford. And on this day a debate was about to begin on the hottest issue of the moment: Darwin's theory of natural selection.

Darwin had published his ideas just seven months earlier, on

November 24, 1859. *On the Origin of Species,* more than twenty years in the making, was an overnight sensation. The entire first printing, all 1,250 copies, sold out to booksellers on the day it appeared. A second edition came out on Boxing Day. Ultimately, six editions of the *Origin* were published during Darwin's lifetime; beginning with the fourth edition, he made substantial revisions, watering down some of his most heretical beliefs to make them more palatable to the Church of England, the biological orthodoxy, and his devout wife, Emma.

Darwin, who had originally studied to become a clergyman, never understood the shrieks of anticlericalism provoked by his theories. "I see no good reasons why the views given in this volume," he wrote in the *Origin,* apparently hoping to preempt the controversy, "should shock the religious feelings of anyone."

But shocked they were. And it seemed that the people with the strongest opinions had convened in this hot library on this last day of June to witness what they knew would be a rousing debate.

On the pro side was, as always, Thomas Henry Huxley, a self-taught anatomist and paleontologist. Darwin himself had no interest in defending his theory. He had neither the taste nor the stamina for confrontation. Writing the *Origin,* a four-hundred-page "abstract" finally finished in a fifteen-month frenzy, so debilitated Darwin — a chronic invalid for most of his life — that he was forced to take a rest cure at Ilkley immediately after the book's publication. He gladly turned over to his great friend Huxley, a member of one of the most illustrious families in England, the responsibility for defending his views.

When Huxley was asked to stand up for Darwin at the British Association meeting, though, his first impulse had been to say no. The association was the preeminent scientific group in the country, uncle to the American Association for the Advancement of Science. And it had assembled an impressive panel of experts to debate the issue. Arguing against Darwinism would be the formidable Bishop Samuel "Soapy Sam" Wilberforce, an amiable, easygoing, witty man, known as a convincing orator despite a rather ordinary intellect. Huxley did not especially relish the prospect of being, as he put it, "Episcopally pounded."

Finally, he was shamed into showing up by Robert Chambers, a prominent essayist, naturalist, author of *Chambers's Encyclopaedia,* and friend of both Darwin and Huxley. But he must have regretted it instantly when he saw how tough this crowd would be. In one dreadful nine-minute stretch, the audience shouted down three speakers. "Let this Point A be Man and that Point B be the monkey," said the last victim, Henry Draper, whose twenty-eight years in America unfortunately led him to pronounce the animal as "mawn-kee."

"Mawn-kee! Mawn-kee!" the crowd roared, dismissing Draper and demanding to hear from the bishop. So Soapy Sam spoke — without much insight into what Darwin even stood for, but with enough sarcasm to ask Huxley one pointed and unforgettable question. Tell me, sir, the bishop asked, is it on your grandmother's or your grandfather's side that you are descended from an ape?

The actual words of Huxley's reply have been muddied by his own and others' later attempts to pretty up what he really said. Among the more colorful contemporaneous accounts — the meeting was widely reported in popular magazines like *The Athenaeum* and *Macmillan's* — a few had Huxley giving such a glib answer that the crowd went wild and one woman fainted. In a letter to a friend, Huxley recalled his response as long-winded and dull: "I would rather have a miserable ape for a grandfather than a man highly endowed by nature and possessed of a great means and influence and yet who employs these faculties and that influence for the mere purpose of introducing ridicule into a grave scientific discussion."

But other accounts offer a more pithy — and memorable — response: "I would much rather be descended from an ape, sir, than a bishop."

Could Darwin have been surprised that his book caused the wild eruption it did? He knew, of course, that it challenged man's position as God's favorite creature, created in his image. That is why he took twenty years to write it, carefully amassing fact after fact to bolster his case. He was finally motivated to publish when a friendly competitor was about to scoop him by going to press with a nearly identical idea. Twenty years after the *Origin* appeared, the religious oligarchy still was condemning it. To call Darwin's idea a "hypothesis," wrote one churchman, was "to do it an honor which it does not deserve. A pile of rubbish is not a palace, and a heap of blunders is not a hypothesis." Even among those who eventually came to accept the notion of "transmutation" — the word then used for what is now called evolution — debates raged about whether species changed gradually or suddenly; about how adaptive characteristics were passed from one generation to the next; and about the precise role and mechanisms of natural selection.

This debate, which persisted into the twentieth century, formed the backdrop for the incorporation of Mendel's paper into the emerging theories of both evolution and genetics. Developments in cell biology in the 1880s and 1890s paved the way for an understanding of Mendel's theories of the discrete factors responsible for inheritance. The renewed interest in Mendel's ideas, in turn, paved the way for an understanding of Darwin's theories about the mechanisms of "descent with modification." Until then, no one could really understand how natural selection worked — not even Darwin himself.

From the moment it appeared, the *Origin* caused great distress among biologists, theologians, and laypeople who believed the book of Genesis was the word of God, its every statement to be taken literally. Among the most distraught were those who followed the teachings of the seventeenth-century scholar John Lightfoot, vice chancellor of Cambridge University, who had declared that he knew exactly when the Creation took place, right down to the moment man appeared: at nine o'clock on Sunday morning, October 23, 4004 B.C.

The development of Darwin's two-part theory explaining transmutation — the underlying "struggle for existence" and the propelling forces of "random variation" and "natural selection" — was itself a study in evolution, one that occurred in small increments over a long period of time. Its first stirrings could be dated back nearly thirty years, to 1831, when the twenty-five-year-old Darwin set sail on the small British exploring vessel the H.M.S. *Beagle,* as a dinner companion to the ship's captain.

The captain, Robert Fitzroy, was afraid of suffering from loneliness and isolation during the long, long voyage to and from the coast of South America. According to Victorian conventions, the captain was not permitted to socialize with his crew, nor even with the physicians, draftsmen, and engineers among the professional men on board. Fitzroy dreaded five years of dining alone and its possible effects on his mental health. The previous captain of the *Beagle,* Pringle Stokes, had shot himself three years earlier during a similar voyage; the burden of solitude had proved more than he could bear. And Fitzroy knew he had a hereditary predisposition to becoming mentally unhinged. Among his ancestors, a long line of aristocrats whom he could trace back directly to King Charles II, were psychotics and suicides, including his uncle, Viscount Castlereagh, who in 1822 had slit his own throat.

That is where Darwin came in. He was perfect as a companion, a young man of the upper class whose interest in the natural sciences helped him see the appeal of sailing to Patagonia. The two men, only a year apart in age, got along well enough when they were on land, and Darwin signed on to the voyage for the sheer adventure of it. But Fitzroy's personality changed altogether once they set sail, and at sea his authority was complete, final, and absolute.

Fitzroy turned out to be insufferable. He held forth at length every night, and Darwin, being little more than a paid conversationalist, could do nothing but listen. All the animals on earth, Fitzroy declared, even the previously unrecorded birds and tortoises found off the South American coast, were fashioned directly by the hand of the Creator. And proof of God's divine plan, of his love for us, of the fact that we are as a species destined to greatness, was the dominance in the British Parliament of the Tory party — a line of discussion that Darwin, a committed Whig, found especially irritating.

How to put up with this for five long years? When Fitzroy's monologues got to be too much, Darwin reminded himself of how rare this opportunity was for a naturalist in training. The ship's itinerary — to the Pacific coastline of South America, including Patagonia, Tierra del Fuego, Chile, Peru, and some of the adjacent islands like the Galápagos — would allow Darwin to create new collections of exotic specimens from halfway around the world, specimens he never would have had access to otherwise. Indeed, his enthusiasm for collecting was so great that within a few months he had displaced the ship's official naturalist, Robert McCormick. Unable to keep up with the pace set by Darwin — who came on board with a servant, a personal fortune, and an amateur's zeal — and occupied by his additional duties as ship's surgeon, McCormick did not have the leisure Darwin had to disembark at every port of call, hire a few willing natives, and go specimen hunting. Nor did he have Darwin's nightly audience with the captain, an important source of power in that setting. In April 1862, after just six months at sea, McCormick was fired; he had to hitch a ride back home on the H.M.S. *Tyne.*

At first Darwin was a staunch believer in the fixity of species.

But he was also aware of the competing theory of transmutation: the ability of species to change and for some species to become extinct. He was introduced to transmutation partly through the works of his grandfather, Erasmus Darwin, who died before Charles was born but whose book *Zoonomia, Laws of Organic Life* was a much-discussed part of family lore. Erasmus Darwin was a vivid man with a scandalous taste for womanizing, who set down some of his thoughts about natural history in the form of erotic poetry, like his 1794 classic, "The Botanic Garden." A devout Christian, Erasmus Darwin believed changes were designed by God and generally led to improvements in species over time. But he also believed that three driving forces were behind transmutation: hunger, security, and lust.

After the *Beagle* returned to England in October 1836, Darwin rented lodgings in London and began thinking about speciation. Mendel was still a boy in Heizendorf, studying at a high school twelve long miles from home, writing poetry about medieval inventors, and dreaming of immortality. Over the next two years, as Mendel moved on to the Gymnasium, even farther from home, Darwin slowly became converted to a belief in transmutation.

From 1836 to 1838 his reading selections were eclectic. He dabbled in geology, a field he was first exposed to aboard the *Beagle*, having set out to sea with the first volume of Charles Lyell's *Principles of Geology: Being an Attempt to Explain the Former Changes of the Earth's Surface by Reference to Causes Now in Operation*. The second volume was sent to meet him when the *Beagle* reached South America. Lyell was of the belief, revolutionary for its time, that geological variations and species extinction took place slowly and continually through the accumulation of a succession of almost imperceptible changes. This went directly against the predominant thinking that change took place rarely and was catastrophic. The way Lyell described it, building on a theory first proposed forty years earlier by a Scottish geologist, the world was in a constant state of flux. By the time he returned to London, Darwin was, like Lyell, a committed "uniformitarian."

Darwin read, too, in zoology and botany, which brought him to the work of Jean Baptiste Pierre Antoine de Monet, chevalier de Lamarck. Today the school of thought named Lamarckism is generally discredited, known and derided for a single idea: the inheritance of acquired characteristics. But Lamarckism also included a theory of organic progression, laid out in Lamarck's most popular work, *Philosophie zoologique,* in 1809. All life arises through spontaneous generation of very simple life forms, said Lamarck, with natural fluids acting on gelatinous matter and "vivifying" it. More complicated life forms arise through transmutation, a steady upward progression in which the nervous fluid carves out more and more complex channels from one generation to the next. Rather than seeing all contemporary life as arising from a common ancestor, however, Lamarck believed that organisms at different levels of complexity arise from different acts of spontaneous generation at different points in time. The higher on the scale an organism is at present, the longer ago its original ancestor appeared, so the more time it has had to develop and progress.

New characteristics, Lamarck said, are acquired according to the "use and disuse theory": nature constantly submits her works to the influence of the environment, and the environment causes changes that affect the electrical or physiological makeup of the tissues. These changes arise not simply from the environment, but from a recognition by the plant or animal of a need for them. Change results from a yearning for change. And, just as important, these changes can be inherited, because they affect the sex cells in a permanent way.

The most famous example of this theory is like one of Rudyard Kipling's *Just So Stories,* one that could be called "How the Giraffe Got Its Long Neck." The story begins when one giraffe in a group, after all the lower, easy-to-reach leaves on a tree have been eaten, perceives the need — stimulated by hunger — to reach the higher leaves. This giraffe yearns for change. It stretches its neck, which increases the flow of fluids to the neck, which makes the neck longer, which further affects the fluids, which further elongates the neck. This longer neck, with its greater flow of cellular fluid, is then passed on to the giraffe's babies. The long-necked babies thrive and in turn pass on their long necks to *their* babies, and on and on through the generations.

Darwin saw an analogy to giraffes in domestic ducks, which have thicker legs and smaller wings than wild ducks because they tend to walk instead of fly. This got him thinking about how environmental differences might account for differences among species.

Then one fall day in 1838, Darwin found his solution by reading an old book from another field entirely. In the forty-year-old treatise "An Essay on the Principle of Population," by the economist Thomas Malthus, Darwin found a phrase that captured his imagination, making all the thoughts that had been roiling his brain and animating his notebooks finally make sense. Malthus wrote about "the struggle for existence."

Yes, existence *is* a struggle, Darwin agreed. Almost every living thing produces more offspring than can possibly survive, given the limitations of the food supply and of a parent's ability to protect its young from predators. Some guiding principle must be involved in determining which of the offspring live and which die. Maybe that was adaptation. Darwin knew there were variations in nature, though he could not explain why. Now, with Malthus's phrase, he could make the next logical leap: that favorable variations tend to be preserved, and unfavorable variations, in the context of the struggle for existence, tend to be destroyed.

Darwin's next step was to offer natural selection as the mechanism that winnowed out the favorable from the unfavorable. He arrived there through an analogy to artificial selection, the breeding of plants and animals. In artificial selection, the intelligence of the breeder pushed the changes in a particular predetermined direction. In natural selection, however, Darwin envisioned no such overriding intelligence. He believed that changes occurred without an end in sight, without any supernatural agent in control — a notion that is said to have turned biology from a rational science to a mechanistic one.

Six years after Darwin's epiphany about the struggle for existence, a wildly popular tract appeared that made him see how careful he would have to be in laying out his ideas about descent with modification. The book, *Vestiges of the Natural History of Creation*, was considered so heretical that its author took great pains to remain anonymous, and his identity stayed a secret until his death twenty-seven years later. This was not an easy secret to keep. The book was immensely popular, with sales of 24,000 in the first ten years after publication. So speculation about *Vestiges*'s author was inevitable. Guesses ranged from the prince consort to Sir Charles Lyell, the geologist. While the book was basically a defense of transmutation, it was presented with a theological spin, with the presumption that species changes, as they unfold, gradually reveal the plan of the Creator. It placed great stock in the notion of a "divine author of nature" moving flora and fauna forward in incremental improvements over time, "the highest and most typical forms always being attained last." According to this theory, all change occurs in a steady, ladderlike progression that inevitably ascends to a higher state.

The year *Vestiges* appeared, Gregor Mendel was making a place for himself in Brünn, spending time in the alpine garden with his new friend Matouš Klácel, attending meetings of the Agricultural Society, getting ready to start classes in ecclesiastical history and archeology at the Brünn Theological College. He was probably not hearing much about the heretical view of transmutation causing such distress in England — not only among theologians but among respected scientists as well. One of the most influential biologists of the time, Adam Sedgwick of Cambridge University (who had, coincidentally, taken Darwin, as a student, on a geological expedition), fired off an eighty-five-page diatribe against the little book. "The world cannot bear to be turned upside down," Sedgwick wrote. "It is our maxim that things must keep their proper places if they are to work together for any good. . . . [O]ur glorious maidens and matrons [must not] poison the strings of joyous thought and modest feelings by listening to the seductions of this author."

In 1871, just after his death, the author of *Vestiges* was revealed to be Robert Chambers, the essayist and naturalist who had prodded Thomas Henry Huxley into his debate with Soapy Sam. By putting the idea of transmutation before the British public fifteen years before the *Origin* appeared, he helped pave the way for an understanding of the bare outlines of evolution long before Darwin came forward to propose the possible mechanisms by which it occurred.

But he also inadvertently made Darwin overly cautious about publishing his own ideas. Aware of the storm unleashed by *Vestiges,* Darwin took his time in publicizing his theory of natural selection, which was more heretical than Chambers's because it presumed no divine plan or ultimate purpose. For Darwin variation was entirely random; the success or failure of a particular adaptation was wholly a matter of chance. He carefully laid the groundwork for the acceptance of his ideas in the scientific community by writing up his theory of descent with modification in 1842. In 1844, the year *Vestiges* appeared, he published the essay at his own expense and distributed it selectively to the men of science with whom he was acquainted and whom he considered most likely to read his work with open and sympathetic minds: Charles Lyell; the botanist Joseph Hooker; and Asa Gray, a botanist from America.

After that he spent his time collecting more and more pieces of evidence to back up his theory. He communicated with plant and animal breeders. He floated seeds, plants, and dead birds to imitate how organisms traveled to distant islands. He enlisted local school-children to gather reptile eggs. He bred pigeons, then killed them to see how their organs might have changed. He did the same with dead ducklings and dead chicks donated by his neighbors. He collected and categorized barnacles — 10,000 of them in all — to draw conclusions regarding the relationship between evolution and Linnaean categorization, which in Darwin's view was a graphic demonstration of the ramifying patterns of common descent. And he did all this from his country home in Downe, which he never again left for more than a day or two at a time. Not only was Darwin bound by the demands of his household, which ultimately included seven children — not to mention several pets and nearly a hundred pigeons — but he was by middle age severely disabled by a strange degenerative condition no one could explain. Retrospective diagnoses have included Chagas' disease, a tropical illness he might have contracted in South America; psychological stress; multiple allergies; or inadvertent poisoning from all the patent medicines he took.

But Darwin did not really have the limitless time he thought he had to piece together his *magnum opus.* A competitor working on a nearly identical theory was ready, in 1858, to announce it to the world.

Alfred Russel Wallace was a professional collector who supported himself by selling specimens gathered on his world travels. Like Darwin, Wallace was first alerted to the variation of species on an expedition to South America. The trip, from 1848 to 1852, had been a revelation to Wallace the way the *Beagle* trip had been to Darwin. But Wallace encountered catastrophe along the way. On the voyage back to England, his ship caught fire, and all his notes and specimens were destroyed. He re-created his notes — and collected his insurance money — and set out again on another journey, this one to the islands of the Malay archipelago in the region now known as Indonesia. On the remote island of Gilolo (also called Halmahera), during a delirium brought on by a tropical fever, Wallace independently conceived of the idea that Darwin was calling natural selection.

Wallace took transmutation as a given. In a brief essay called "On the Tendency of Varieties to Depart Indefinitely from the Original Type," he outlined the mechanisms through which he thought transmutation might occur. "The life of wild animals is a struggle for existence," he wrote. "Those which are best adapted to obtain a regular supply of food, and to defend themselves against the attacks of their enemies and the vicissitudes of the seasons, must necessarily obtain and preserve a superiority in population." In June 1858 he sent a prepublication copy to the man he knew could best appreciate it: Charles Darwin.

Wallace had known Darwin for three years, though the two had never met. In 1855 Wallace published a paper stating that "every species comes into existence coincident in time and space with a preexisting closely allied species." Darwin wrote to express his agreement, and the two began a lively exchange; Darwin was a fervent letter writer. During all this time, though, the older man never intimated that he was working on a theory based on an assumption quite similar to Wallace's.

And now here was a fully fleshed-out theory that looked almost exactly like Darwin's own. Darwin spun into a panic. For twenty years he had arisen from his sickbed for a few hours every day to work on descent with modification. Now he could see the error of moving too cautiously. He had been scooped — by a mere specimen hunter.

Darwin's friends decided to stake out a priority claim on his behalf to ensure that Wallace could not make one himself. "I was at first unwilling to consent," Darwin said in his autobiography, "as I thought Mr. Wallace might consider my doing so unjustifiable." But consent he did, and as there was already ample evidence that Darwin had arrived at his own theory by 1842 — especially in the essay that he had distributed to a few well-placed biologists — he did not anticipate much difficulty convincing the scientific community that he and Wallace had, quite independently, arrived at similar conclusions along parallel roads. (Nonetheless, occasional charges of plagiarism have clouded Darwin's reputation even to this day.) On July 1, 1858, Darwin's friends Lyell and Hooker read three papers at a meeting of the Linnaean Society in London: Wallace's article plus extracts from Darwin's 1844 essay and a letter describing his theory that he had written to Asa Gray on September 5, 1857 — before reading Wallace's report.

In a strange foreshadowing of the reception that would be granted to a similarly revolutionary paper, the one Gregor Mendel would deliver seven years later, Darwin's and Wallace's contributions were all but ignored. Perhaps because, like Mendel, they were too far beyond the conventional thinking of the day, no one in the audience at the Linnaean Society meeting asked any questions, and no one took much notice of the event. The papers appeared side by side in the society's *Proceedings*. But the tandem publication, like Mendel's, caused hardly a ripple. Darwin recalled just one written reply to the articles, from an Irishman who said "all that was new in them was false, and what was true was old."

After this, Darwin set to work. He had been gathering information from plant and animal breeders, who created new forms of the species they cultivated. But what were these new forms exactly? New strains? New varieties? Entirely new species? And what was the mechanism through which such crossbreeding worked?

What he needed most emphatically was a theory of inheritance; his conception of natural selection was only half finished without one. As he saw it, environmental forces helped perpetuate those variations that provided what he called a "selective advantage," a benefit to the plant or animal that allowed it to produce offspring in greater numbers than would a competing organism without that variation. But how exactly was such a trait passed on? And why did it persist?

This is where Mendel could have helped — had either he or Darwin been able to see clearly how random rearrangement of the units of inheritance explained the variations needed to propel natural selection. In the absence of any knowledge of Mendel, though, Darwin was seriously hobbled. Having no facility with numbers, he was hopelessly confused by most other notions of inheritance then being proposed. Indeed, his mathematical obtuseness often tested the patience of anyone who tried to explain inheritance to him, especially his son George and his cousin, Francis Galton. When Galton, one of Darwin's closest friends, devised the "law of ancestral inheritance," the numbers made Darwin's usually brilliant brain turn to porridge. This law assigned a fraction to represent the proportion of inherited traits that come from each of a child's two parents, then from each of his four grandparents, each of his eight great-grandparents, and so on. According to the law, each parent (designated as *p*) gives a child one-fourth of his genetic complement. Each grandparent *(pp)* gives one-eighth; each great-grandparent *(ppp)* one-sixteenth; each great-great-grandparent *(pppp)* one-thirty-second. In this way, a trait in an individual's lineage is never lost; it is only diluted.

Darwin did not pursue Galton's theory or anyone else's, but instead came up with his own — one involving no numerical calculations at all. He called it "pangenesis," and said it involved units he called "gemmules." Gemmules, generated by body cells, travel to the gametes through the bloodstream or, in the case of plants, through a system of internal transport known as phloem channels. There the gemmules wait, in a dormant state, until the moment of fertilization, when they are transmitted to a new generation. Since gemmules originate in body cells, Darwin saw them as a mechanism for the inheritance of acquired characteristics. Something in the environment causes an organism's gemmules to change, and the gemmules travel to the gametes and pass on their changes to the plant or animal's heirs.

For Darwin, gemmules were also the mechanism for blending, his favorite theory of inheritance. But if traits blended, his opponents said, wouldn't a new variation quickly be blended into nonexistence? As one of Darwin's harshest critics, the physicist Fleeming Jenkin, put it, when a rare mutation (he called it a "sport") occurs, it would soon be wiped out, because the sport, as the only one of its kind around, would be forced to mate with a normal individual, and their offspring would "on the whole be intermediate between the average individual and the sport." The result, Jenkin said, was the rapid elimination, or swamping, of a randomly generated mutation from the general pool of variation — like that single drop of red paint in a gallon of white, the drop that, after a few good stirs, completely disappears.

But swamping would be no problem, Darwin replied, if the conditions of life kept changing. He always considered changes in the environment to be the primary source of variation; the most highly adaptive changes were those that persisted. If a bear living in a cold climate gradually laid down more and more fat, making it more likely than thinner bears to survive the winter, then its cubs would be born already fatter than most. This notion of the inheritance of acquired traits was one of Darwin's most abiding beliefs, one of his most persistent blind spots in his search for mechanisms to explain how natural selection worked.

He refined this idea with another notion, the increasing hereditary strength of traits over time, that has also since been discredited. He called it "Yarrell's law" after his friend William Yarrell, a newspaper wholesaler whose free time was occupied with the "country passions" of hunting, collecting, and breeding animals. According to Yarrell's law, the oldest traits are the strongest, and are more likely to be inherited than are traits only recently introduced into a species. This would emphasize nature's conservative tendency: whatever blending might occur between an old variant and a new sport would lean in the direction of the older, stronger characteristic.

Debates with men like Fleeming Jenkin occupied Darwin's supporters for decades. The showdown between Huxley and Soapy Sam was replayed again and again in overstuffed lecture rooms, stifling courthouses, and, even today, cool chambers of state school boards deciding whether evolution is a suitable subject for the science curriculum. When the rancor began, in the 1860s, one of the most dramatic defenders of Darwin's theories was Francis Galton.

To Galton, a brilliant mathematician, biologist, and statistician, anything could be quantified: the efficacy of prayer, the relative beauty of women, the girth of the British peerage over three generations. Galton was one of the last of the gentleman-scientists, pursuing ideas that charmed him just for the fun of it. He took a dilettante's approach to his studies, too, deciding while at medical school in London, at the age of fifteen, to work his way through his pharmaceutical textbook by sampling every drug listed there. Tackling the job alphabetically, he only made it as far as the C's, undone by the purgative effects of croton oil. With little else in the curriculum to hold his interest, he soon abandoned medicine.

Galton dabbled in just about everything. He discovered that fingerprints are unique. He was the first to name and describe the anticyclone. And in the 1860s he spent a great deal of time trying to help Darwin understand the questions of inheritance that dogged him throughout his life.

Galton was not a blind follower of his more illustrious cousin; indeed, one of his first experiments was designed to show why

Darwin's pangenesis theory was mistaken. Beginning with groups of rabbits with different-colored coats, Galton transfused the blood across coat colors: a white-furred rabbit got blood from a brown rabbit; a brown rabbit got blood from a white. According to Darwin's theory, gemmules would be transmitted along with the blood, which should have turned the white rabbits brown and vice versa. But after Galton bred them, he found that the blood transfusions had no effect at all on the coat color of the offspring. White still bred white, brown bred brown.

Galton went on to mount explorations to uncharted territories in southwest Africa, to originate the statistical concepts of regression and correlation, and to coin the word "eugenics" — a social application of genetics to human breeding that has since justified some of the most brutal episodes of racial purification, ethnic cleansing, and genocide in recent history. Involved as Galton was in what have become some of the most vociferous debates of twentieth-century science, it is a surprise to learn that he was born in 1822 — the same year as Gregor Mendel. The two men seem to come from different generations.

Blessed by great longevity — he died in 1911, at the age of eighty-nine — Galton was witness to the early days of the new science of genetics, when it was at its stormiest and, arguably, most interesting. The two sides that fought most bitterly during the early twentieth century, especially in England, claimed to be the standard-bearers for, respectively, Darwin on the one hand and Mendel on the other. But each side also laid claim to a particular leader and inspiration: each one claimed Galton.

Galton did indeed inspire both the ones who called themselves Darwinians, and the ones who called themselves Mendelians. But when Mendel himself was at work, during the early 1860s, Galton was unaware of him. At that time Darwin was stirring up great swells of hysteria throughout Britain, the Continent, and America, and Galton devoted himself to explaining Darwin to his peers and explaining his peers to Darwin. All during this time, Mendel — Galton's exact contemporary — was creating a storm of his own of an entirely different sort. Mendel's storm raged mostly inside his head as he walked among green and growing vegetables in a quiet Moravian garden. And it would not erupt out loud for another forty years.

10

Garden Reflections

Leaves begin to yellow and brown. Flowers become seeds. Everything is soft, large, ripe. As I walk among the plants, they reflect my mood — placid and self-satisfied.

— *The Undaunted Garden*, Lauren Springer

THE AUTUMN GARDEN turned indigo as Mendel raced to collect the last of the pea pods. Just a short while remained until darkness, and he was in a hurry — a situation he found irksome. More his style was to walk slowly among the plants, which he liked to call his "children" to get a reaction from visitors who did not know about his gardening experiments. "Would you like to see my children?" the priest would ask. Their startled and embarrassed faces were always good for a little chuckle.

If you heard Gregor Mendel tell the story, you might think that on this late afternoon in the autumn of 1862, he still did not fully understand the truths about inheritance that his surprising little children — and their relationships to their parents, siblings, and offspring — had so cleverly concealed. Like any good parent, he had hovered over his pea plants for six long years, harvesting the pods, picking out the peas, dividing them into piles according to color or shape, keeping track of the height or flower color or blooming habit each plant displayed at maturity. But, like any parent, he still did not really understand his own children. Yes, he had spent his idle hours during those six years turning over in his mind

Pisum's quirks and peculiarities, the way a parent might dote on and fret over a beloved brood. But the scientific significance of all these observations was only now beginning to dawn on him.

This, at least, is how he would eventually relate, in his best teacherly fashion, his gradual understanding of the laws of inheritance. In his telling, his thought process seems more sterile and straightforward than it probably was. Like other scientists before and since, Mendel imposed a narrative on his *Pisum* research that did not convey its true excitement. The story gained something in terms of clarity, as though his thinking were so logical and clean as to have been almost inevitable. But what he lost was probably more important: the chance to convey the truer, more compelling nature of experimental work as something mysterious, meandering, and messy.

So Mendel's reconstruction of his research probably stretched the truth a bit — and in the process sold short his own insight and foresight. Given his familiarity with physics and hypothesis-testing, it is hard to believe he would not have known at each stage what to expect of his peas, and why. No doubt as the years went on and his results dribbled in, some of his hypotheses changed. But it was disingenuous of him to claim, and mistaken of us to believe, that he gardened first and thought things through second.

His thinking took place outdoors in the garden or indoors, as he sat in an armchair in the library's musty recesses with a book open in his lap. Sometimes he sat thinking in the orangery; at other times he climbed the stone steps behind the orangery to the next level of the monastery grounds, where his bee house was. As his pea experiments required less gardening and more reflection, Mendel spent more time with the bees, renewing a hobby from his boyhood — beekeeping and honeymaking. Years later he tried to confirm some of his plant results by breeding bees.

With age, however, Mendel was less likely to climb the steps to the bee house, finding himself, as he put it, "blessed . . . with an excess of avoirdupois [that makes] hill-climbing very difficult for me in a world where universal gravitation prevails." But the vista alone was worth the struggle of the climb. The red tile roofs of the monastery, the steeple of the basilica, the looming hill on which the Spielberg castle rose, the bustling Brünn streets that fanned out from the Klosterplatz — the view gave Mendel a chance to think, to rest his weakening eyes, and, on his worst days, to renew his sense of mission and remember how important it was to keep up his work in the garden, no matter how frustrating or tedious it might be.

Until 1862 the view from the monastery hill was about as broad a vista as Mendel ever saw. Not counting his trips back home to Heizendorf or his forays to Vienna — some of which were wonderful and exciting, some tinged with bitter disappointment — Mendel was essentially homebound for his first forty years. But in the summer of 1862 he took an extended trip abroad that gave him a taste for travel that lasted the rest of his life. Along with other teachers and the headmaster of the Realschule, Josef Auspitz, Mendel was part of the official Brünn delegation to the first annual London International Exhibition, a technological extravaganza for which the Realschule had prepared a display on crystallography.

The exhibition was mammoth. It followed by a decade the great London Exhibition of 1851, which had been history's first world's fair. The earlier fair's most lasting legacy was the splendid Crystal Palace, a huge domed edifice of steel and glass, an architect's marvel because it was the first prefabricated structure of its kind. The 1862 exhibition emphasized technology rather than artistry, and everything about it was done on the grandest of grand scales. The exhibition hall, though not nearly as beautiful as the Crystal Palace, was the largest ever built, measuring 1,200 feet long and 560 feet wide and covering sixteen acres. Nearly ten thousand individuals applied for exhibit space, at least seven times the number that even this huge building could hold. The most fanciful proposals came from amateurs: perpetual motion machines; the oldest piece of bread in the world (from 1801); epic poetry to be hung in the picture gallery; spring-loaded boots; a mustache guard for eating soup; and the embalmed body of one Julia Pastrana, a bearded Mexican woman who had been exhibited, in one of the many freak shows that sprang up after the *Origin of Species* appeared, as "part human, part orangutan." An accountant in the City, where the most buttoned-down of London's businessmen could be found, offered three decidedly nonbuttoned-down inventions: a self-activated water closet, an improved theodolite (a surveying instrument that measures angles both horizontally and vertically), and an "omnitonic flute" capable of playing every note the human ear could hear. A bookbinder offered a plan for a suspension mechanism for bridges and viaducts; an insurance broker, a better way to make wine. A nurseryman proposed improvements in surgical instruments, while a surgeon offered a wall-hung contrivance to speed the ripening of fruit.

The official country-by-country exhibitions were staid by comparison. The man who wanted to fly in the 260-foot-high glass dome was turned away, but the delegations from the various nations could take up space with dozens of flat-footed exhibits touting their biggest industries. Austria displayed its merino wool, leather goods, stearic acid candles, dyed fabrics, and elaborately carved smoking pipes, as well as a mountain-climbing locomotive. It offered a demonstration of the process for removing tin from discarded iron plate, and another showing the range of food, fabric, and paper products that could be made from maize. Similar litanies of promotion characterized the exhibits of each of the two dozen countries represented.

Predictable though the London Exhibition might have been, city officials from Brünn saw it as a rare opportunity. Brünn was a center of industrial activity (primarily textile manufacturing) with a vibrant economy. The town fathers believed the best way to continue this upward spiral was to invest in projects with more and more technical sophistication. One of their ideas was to build a new Technological Museum — and the London Exhibition could show them, or their emissaries, how to design such a place.

As Mendel helped create the Realschule's display, did he realize that the logic of crystallography could help him make sense of the ratios he had compiled from his peas? Did his work with crystals lead him to the generalization that nature is based on discrete characters? Did crystals somehow form the template for Mendel's emerging understanding of his dizzying collection of numbers — of peas, of pods, of plants — that he was trying to assemble into some sort of algebraic expression that would make sense of inheritance?

Like the units involved in heredity, the crystals Mendel was studying replicated themselves. In both peas and crystals, particles that seemed to be totally inert proved capable of carrying out one of the most central functions of living things. The shape of the existing crystals defined the shape that a new crystal could assume, restricting the possibilities by acting as a template. In the same way, although Mendel probably did not know it yet, the determinants of the traits he was studying in his peas restricted their possibilities, too. A pea could be either round or angular, not in between; tall or dwarf and nothing else. In some significant but as yet mysterious way, the determinants kept everything within circumscribed bounds, just as the crystal structures did, limiting the possibilities of how different from its ancestors a thing could become.

"To London, to London, to visit the Queen," goes the old nursery rhyme, and Mendel may have been every bit as charmed by his trip abroad as was the narrator of the poem. Few documents from the excursion remain, but we do have a photograph of the entire Moravian group, more than thirty strong, standing in front of the

Grand Hotel in Paris during one of their stops before crossing the English Channel. Near the hotel the resplendent Paris Opera would soon be erected. The photo shows nearly three dozen men and one lone woman posing on the hotel steps, palm fronds drooping from above. Most of the men in the photo are unidentified. But here is Mendel, standing in the precise middle of the group and looking off somewhere past the photographer's left shoulder. The only person with spectacles — and nearly the only one without a bushy mustache or full, rounded beard — Mendel does not give himself away as a priest. He is wearing the same sort of morning jacket as everyone else, with a white shirt and a dark cravat. He is not smiling. And even though a few of the men stand leaning against one another or with an arm draped across a neighbor's shoulder, Mendel stands erect and alone.

The group had first stopped in Vienna, the city Mendel had come to know so well as a student. How many memories the visit must have evoked, both of the euphoric time spent soaking up new information on every scientific subject he could, and of the despairing times when he failed — and failed again — to pass the only examination he ever cared about. A handful of German cities were also on the delegation's route to London: Salzburg, Munich, Stuttgart, Karlsruhe, and Strasbourg.

The trip took about three weeks in all, from July 24 until the middle of August. Some have suggested that Mendel or his countrymen could have met Charles Darwin during this visit. How gratifying it would be to imagine such an encounter. "I do not understand the mechanism for the natural selection process," Darwin might confess, speaking in his stilted German or using an interpreter. (Mendel, who was raised speaking a Silesian dialect of German, wrote in formal German and taught classes in Czech. He spoke not a word of English.)

"Oh, yes, I could tell as much in your *Origin*," Mendel might answer. He had read the book in German translation, *Über die*

Entstehung der Arten, as soon as it appeared in 1860, and had heard much about the controversy it caused through conversations at the monastery and lectures at the Natural Science Society. He consumed Darwin's book, penciling notes in the margin in his small and careful handwriting, revealing the intensity of his feelings with an occasional double underline or even an exclamation point. Though Mendel agreed with Darwin in many respects, he disagreed about the underlying rationale of evolution. Darwin, like most of his contemporaries, saw evolution as a linear process, one that always led to some sort of better product. He did not define "better" in a religious way — to him, a more evolved animal was no closer to God than a less evolved one, an ape no morally better than a squirrel — but in an adaptive way. The ladder that evolving creatures climbed led to greater adaptation to the changing world.

If Mendel believed in evolution — and whether he did remains a matter of much debate — it was an evolution that occurred within a finite system. The very observation that a particular character trait could be expressed in one of two opposing ways — round pea versus angular, tall plant versus dwarf — implied limits. Darwin's evolution was entirely open-ended; Mendel's, as any good gardener of the time could see, was closed. And neither of them, significantly, saw the hand of God in the process of species change.

Like the geneticists who would follow in his footsteps, Mendel doubted Darwin's belief in blending inheritance. "I don't think blending can serve as an explanation," he might say in our imagined meeting between the two scientists in London. "According to the laws I'm uncovering, at least the laws as they apply to *Pisum,* traits do not blend. They remain separate and are passed on independently."

"I'm rather relieved to hear you have an alternative explanation," Darwin might admit. "I'm tired of hearing from my critics about the swamping effect. So tell me — what do you know about how adaptations might be preserved?"

But it is highly unlikely that such a meeting did take place. Even granting the farfetched possibility that Mendel was bold enough to request a meeting with the most prominent biologist of his day, other, more pressing events would have prevented an encounter. While the Brünn delegation was in town, Darwin's twelve-year-old son, Leonard, was seriously ill with scarlet fever, and his parents stayed with him at their house in Down, receiving no visitors. It must have been a terrible vigil for the Darwins, who by then had already lost three children, two in infancy and one, ten-year-old Annie, just the year before. But Leonard survived — and lived on to the ripe age of ninety-three, the only one of Darwin's ten children to see his father's work become the basis for an entirely new way of thinking, not only for biology but for all of modern science.

When his London excursion ended in the fall, Mendel returned to his garden. He went back to tending his peas and working through his calculations — the ones he called the most difficult in his experiments so far.

Mendel's monohybrid and dihybrid crosses had by this time been carried out to at least the third generation (that is, what we would now call the F4's). Some of these early crosses were still growing in the garden, their lineages recorded as far as generation six. Now, on this darkening autumn afternoon in 1862, Mendel was finishing up the fourth generation of his most ambitious cross, the trihybrid. This was a cross between dominant and recessive parents that differed along not just one trait or even two, as his earlier monohybrid and dihybrid crosses had. These were crosses between plants that differed in three distinct characteristics: pea shape, pea color, and seed coat color.

Why did Mendel choose this particular trait, seed coat color, to add to the other two? This is a true mystery, because his choice created several new problems. First, the seed coat was a characteristic not of the pea — which was the offspring of the plant on which it grew — but of the plant, a characteristic that did not show up until the spring after the peas emerged in the fall. This meant there would be a built-in hiccup in Mendel's results: he would know about the first two components of his trihybrid cross (pea shape and color) more than nine months before he found out about the third.

In addition, it was impossible to see the color of the pea without first splitting open its coat. This meant Mendel had to keep especially careful records and had to write down his findings right in the midst of manipulating his peas, which were difficult enough for the thick-fingered, short-sighted Mendel to work with even without stopping in the middle to write things down. Maybe this was one of the ways in which he made use of Alipius Winkelmayer, a fellow monk, who occasionally showed up in the garden or the orangery to ask, in his genial way, whether he could be of any help.

After the complicated trihybrid cross, Mendel still needed to perform one more experiment. This final test of all his theories and deductions would require great care, possibly proving even more difficult than the trihybrid crosses. But he could not trust his earlier results until he had gone through with it. Other hybridizers used this approach, usually without proving anything conclusive. Somehow, Mendel believed his results would be different.

The technique was known as a backcross, or a test cross. It involved crossing some of the F1 hybrids with purebreds — either pure dominants or pure recessives. Because Mendel used dihybrids in his backcrosses, the purebreds were either double dominants *(AB)* or double recessives *(ab)*. In this case he chose peas that differed according to shape and color. The double dominants were round and yellow, the double recessives angular and green.

The F1 double hybrid peas *(AaBb)* with which Mendel began his backcross looked round *(Aa)* and yellow *(Bb)*, just like the double dominants. Despite their appearance, though, Mendel's hypothesis was that the hybrids carried within them determinants from their recessive parents, determinants for angular and green. He deduced the presence of these recessive traits after coming up with his 3:1 ratio of dominants to recessives among the hybrids' F2 progeny. The existence of that one green and angular pea for every three that were yellow and round seemed to confirm Mendel's hypothesis about dominance and recessiveness, and about traits being passed on to offspring consistently, randomly, and independently. Now a backcross would offer even better proof.

In the spring of 1862, before he left for London, the backcrosses began. Mendel crossed half his double hybrids (the ones we are calling the F1's, designated as *AaBb*) with pollen from double dominants (*AB* to Mendel, or what we would now write as *AABB*) and half with pollen from double recessives (*ab*, or, in modern shorthand, *aabb*). Then he closed up the flowers, protecting his backcrossed hybrid embryos with the same kind of caps he had been festooning his flowers with for six years, and waited.

Now, in late summer, it was time to harvest the F1 pods and look inside to see what the peas' shape and color told him about the next generation.

According to Mendel's hypothesis, dihybrids should produce four kinds of gametes in equal proportions: *AB, Ab, aB,* and *ab*. When crossed with the double-dominant parent, the offspring should all look the same, no matter which hybrid gamete had been involved. This is because in every case the *AB* gamete from the double dominant — the only kind it could produce — would hide whatever recessive trait a hybrid might be harboring.

A double-recessive backcross, in contrast, should lead to four types of progeny, each one different in appearance. The appearance, moreover, should reveal immediately which hybrid gamete was involved in its formation, because the purebred parent contains none of the dominant gametes that would mask the true nature of the F2 hybrid.

In the recessive backcross, Mendel reasoned, each of the four

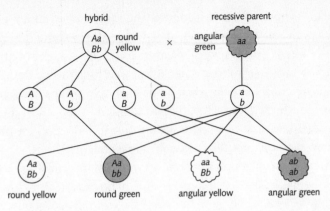

A double-recessive backcross of round yellow and angular green peas.

types of hybrid gametes should produce a pea that looks different from any of the other types. And each of the types should have an equal chance of occurring. If the *ab* from the double recessive fertilized the *AB* gamete from the hybrid, you get *AaBb*, a pea that is yellow and round. If *ab* fertilized *Ab*, you get *aAbb*, yellow and angular. Cross *ab* with *aB* and you get *aabB*, green and round. And cross *ab* with *ab* and you get another double recessive, a pea *(aabb)* that is green and angular.

This is exactly what happened. Of the 203 peas produced by the double-recessive backcross, Mendel counted 55 that were yellow and round, 44 yellow and angular, 51 green and round, and 53 green and angular. The ratio among the four types was nearly 1:1:1:1.

This was what Mendel had been looking for. "The crop fulfilled these expectations perfectly," Mendel said, barely able to contain his excitement. "There could scarcely be now any doubt of the success of the experiment."

With this declaration, Mendel took the first steps toward the foundation of modern genetics by demonstrating, at least if one reads between the lines, that he understood the difference between a plant's appearance and its underlying makeup. He did not have the vocabulary to explain it; that would not come along until nearly fifty years later, when scientists understood more about the cell, the nucleus, and the gene. But he was nonetheless revealing the distinction between what we now call phenotype and genotype. To Mendel, it was the logical extension of the law of dominance. Because dominating traits are able to hide recessive ones, not every round or yellow pea is the same underneath.

Mendel carried out his backcross for one more generation to be sure he had really confirmed his own hypothesis. The following spring he planned to sow the seeds he was now collecting and counting, and when the time came he would let them self-fertilize. By the fall of 1863, he would have his F3 generation from both types of backcrosses, the double dominant and the double recessive. He expected the 3:1 ratio to emerge from the half of the peas that were true hybrids. The others, the double dominants and double recessives, would both always breed true, next year and forever after. In twelve more months, then, he would find out if his prediction about the F3 generation was correct.

Pea growing occupied Mendel only for another year or two: by 1863 all his experiments were complete. And the following spring the nefarious pea weevil, *Bruchus pisi,* essentially wiped out the crop. By then, however, Mendel had moved on to other species, including beans, snapdragons, sweet William, and maize — plants that in his estimation generally confirmed his conclusions with peas. This wider experience led him to believe that "the law of development discovered for *Pisum* applies also to the hybrids of other plants."

We can only speculate about what really happened. We do not know exactly how the experiments were done, in what order, during which seasons, even precisely where in the wide courtyard of the St. Thomas monastery in Brünn. We do not know for sure how many generations Mendel squeezed into a single growing season, nor how often he grew plants in the greenhouse and how often in the garden. Nor do we know the total number of pea plants he used, whether anyone helped him in his labors, or where he was on any particular day during the most intense period of his experimentation.

Mendel seems not to have kept lab notebooks, or, if he did, they were later destroyed. All we know for sure is the way he chose to describe his work in two public lectures and in a handful of letters to a German botanist.

But what if he did his research differently from the way he chose to relate it? What if he began, not with monohybrid crosses — painstakingly repeating the process in the second generation, the third generation, the fourth, and in some cases the fifth, sixth, and seventh — examining his peas for green versus yellow, round shape versus angular, for six years, till his eyes were about to burst and his head spin with the tedium? What if he began by doing everything all at once, and only later — in 1860, perhaps — separated his findings to see what happened to a single trait at a time? What if he later reconstructed his experimental design to make it seem as if it had taken place in the exact reverse order — pretending he had moved from the simplest crosses to the hardest instead of the other way round — to make his audience see the logic more clearly?

Mendel could almost as easily have started with a complex, multilayered set as the way he said he had proceeded. For a variety of reasons, which would not become an issue until forty years or more after his paper's rediscovery, a reverse scenario, in which individual traits are winnowed out through a method called the forked-line approach, makes sense scientifically. It also answers some of the charges that haunt Mendel's reputation even to this day: that he molded his data to make his 3:1 ratio more convincing, that it is suspicious that he never encountered linkage and the other confounding genetic phenomena that would have cast doubt on his earliest observations. With the forked-line method, precise ratios are more the rule than the exception, and oddities like linkage are less likely to emerge and spoil the picture.

But it is difficult to settle, a century and a half after the fact, even so basic a matter as what happened first and what second. All we really have to go on in reconstructing Mendel's life's work is a single publication, which Mendel himself called a summary of his public lectures — and which was probably written to make things clear rather than accurate. How could he know that today his little forty-four-page paper would still be picked apart and analyzed with Talmudic thoroughness? He thought he was giving the equivalent of a biology class to a group of forty self-taught but enthusiastic naturalists. It turns out he was putting together his direct line to the future, creating words where few existed to provide an insight into the origins of the most revolutionary science of our time.

His presentations may not have been clear enough, however. Judging from the reaction of his audience, and of almost every one of his contemporaries who subsequently read the lectures in published form, no one fully understood what Gregor Mendel was trying to say.

II ✦

Full Moon in February

*Nothing here is in a hurry. There is no rush toward
accomplishment, no blowing of trumpets. Here
is the great mystery of life and growth. Everything
is changing, growing, aiming at something, but
silently, unboastfully, taking its time.*

— *How to Have a Green Thumb Without an
Aching Back*, Ruth Stout, 1884–1980

WINTERS WERE ALWAYS DIFFICULT for Mendel. Prone to melancholy, he found the long nights and anemic days especially trying, craving as he did the balm of light, warmth, and sunshine. He was susceptible to chills, so the harsh weather of Brünn was not only unpleasant for him but risky. If he could have managed it, he would have spent the season hibernating.

Second best was to pass the time with the tropical plants that were, in a sense, already in hibernation in the orangery. There he would sit and study and meet his young nephews, who attended Gymnasium in Brünn, for a Sunday afternoon game of chess. But even in the warm, sweet-smelling orangery, surrounded by potted plants and trees, Mendel's spirits sagged. He spent as much time as he could in the monastery library, took pleasure in such indoor activities as reading and eating, and smoked up to twenty cigars a day, a habit his doctor recommended for weight loss. But none of these diversions was as good for his soul, or his rapidly aging body, as the time he spent in other seasons in his garden with his "wards," his "children," the delicate yet resilient growing things he so dearly loved.

Perhaps it was no coincidence, then, that Mendel assigned himself the task of collecting the data from his completed series of experiments and presenting them to his peers in the dead of winter. The timing was fortuitous. His pea experiments were finished, he believed that his work with other species had generally confirmed his findings, and he felt ready to take the next step in a scientific undertaking: stimulating others to try to replicate his results.

Focusing on the writing of his lectures kept Mendel happily occupied through the long dark afternoons of the winter of 1864–65. All that meticulous, backbreaking work. All those hours spent thinking about peas — peas in the garden, peas in the greenhouse, peas in the library and the orangery. All that time with tasks that were occasionally thrilling but far more often tedious. All that effort, eight years' worth, seemed at last about to pay off.

On Friday, February 8, he was ready to make his presentation. The weather, though frigid, had turned delightfully clear. Under a crystalline sky, Mendel and a few fellow monks walked up the winding hill of Bäckergasse toward the Realschule, where Mendel had been teaching physics and natural science for eleven years. It was a resplendent stone edifice in the style of a Florentine palace, still relatively new in 1865, located a few blocks past the central Cabbage Market, just inside the city gate. Its bow windows and the tall corner clock tower took on an eerie luster at night, glowing under the gaslights of the Johannesgasse and the steady shining of a full February moon — the "hunger moon."

Despite his girth, Mendel looked almost dashing in his long black coat, high boots, and tall black hat. Under his arm he carried his handwritten manuscript and a few samples of the pea plants he was about to discuss. As he arrived in the hall, Mendel was greeted by his friend Gustav von Niessl, a professor of astronomy and botany who served as secretary of the Brünn Society for the Study of

Natural Sciences, sponsor of the evening's lecture. Like hundreds of similar groups scattered across the Continent, the Brünn Society comprised both professional scientists (like Niessl) who taught at the local Gymnasia, technical colleges, and universities, and amateur naturalists (like Mendel) with little formal training but an avid interest in self-education and debate.

On this night the audience numbered about forty. Most of the faces were familiar to Mendel. There was the noted botanist Alexander Makowsky, who taught with him at the Realschule, as well as the chemist Franz Czermak. Jacob Kalmus, a physician, was there, and Mendel's fellow monk Antonin Alt. Also in the audience were some men Mendel knew from his student days in Vienna: Karl Schwippel, a geologist, who by then was teaching at the Brünn Gymnasium; and Joseph Sapetza, a mineralogist.

But missing was Johann Nave, who had been one of Mendel's closest friends in Vienna and, later on, in Brünn. After Nave's move to Brünn in 1854, he remained enamored of the natural sciences. The two men often spoke about their botanical obsessions, Mendel's with peas and Nave's with algae. Nave had died the previous year, at only thirty-four; it had fallen to Mendel to administer his friend's last rites. Had Nave lived to see this evening, he might have been one of the few men in the audience to understand the significance of what Mendel was about to say.

As the murmuring men shuffled to their seats, the florist Karl Theimer rose to the podium. Theimer, another friend from Mendel's university days, was by training a pharmacist. It was his duty, as vice president of the society, to preside over the meeting on this cold February night. After going over some society business, he introduced Mendel, and the audience grew still.

The monk took a deep breath and began. He was nervous, no doubt; a soft-spoken man, he was shy and unaccustomed to public speaking, even in front of a small group of people he knew quite well. He read his paper — which was filled with mathematical ratios and modest in its claims — for about an hour. There were no questions. The shuffling began again, and the meeting broke up. Precisely four weeks later, in early March, Mendel read the paper's second half; again the response was courteous but quiet. No one asked a single question, because no one understood the significance of what Mendel had discovered. For all they knew, they had just spent two boring evenings listening to a local friar describe his gardening work.

That, at least, is the standard version of the lectures' reception. A different one can be gleaned from the articles in Brünn's daily newspaper, *Tagesbote*. Because Mendel's superior at the Brünn Realschule, headmaster Josef Auspitz, was on *Tagesbote's* editorial board, the accounts of the February 8 and March 8 lectures were perhaps more insightful than they might have been otherwise. Luckily, this anonymous reporter — perhaps Auspitz or Niessl — understood enough to produce a respectable summary of Mendel's paper. He described the audience reaction as anything but bored. "The numerical data with regard to the occurrence of the differentiating characters in the hybrids and their relation to the stem species were worthy of consideration," the reporter wrote. "That the theme of the lecture was well chosen and the exposition of it entirely satisfactory was shown by the lively participation of the audience."

It is also possible that the *Tagesbote* article was written by Mendel himself. Following a convention of the time, Auspitz might have asked his colleague to summarize his own report, writing the newspaper accounts as though the lectures — and their anticipated reception — had already taken place. This could mean that Mendel never actually saw any "lively participation" on the part of his audience, simply that he had hoped to.

Two years later Mendel wrote a letter in which he referred to the "divided opinion" he had encountered following his first lecture. But this phrase, while it implied that there had been at least some discussion, did not necessarily mean that anyone in his audience had understood enough of the lecture to have asked a question. Niessl, for instance, listened to Mendel quite politely, but despite his prominence and broad education he, like the others, had little to say.

During the first hour's talk, in February, Mendel presented the results of his experimentation over the previous eight years. He described his 3:1 ratio, and the subsequent self-fertilizations through succeeding generations that had led him to refine the ratio to 1:2:1 (one pure dominant for every two hybrids for every one pure recessive). He explained why this ratio could be expressed algebraically as a combination series, an expression of terms linked by plus signs. Each parent's contribution is represented by letters, which are multiplied together to derive the combination series describing the offspring (or, in algebra, the product). If each parent is *Aa* — in other words, hybrid for the characteristic symbolized by *A* — the product of *(Aa)* × *(Aa)* is the combination series *A + 2Aa + a*; that is, one dominant offspring for every two hybrids for every one recessive.

The simplicity of the math belied the arduous process of arriving at this expression. Most of the work took place during long winter evenings, though no one knows for sure exactly which winter it was. The best guess is that Mendel's mathematical insights occurred in a burst sometime between 1859, when his first round of F3 hybrids produced their seeds, and 1862, when the bulk of the growing was complete. During these months the priest covered page after page with calculations about how a single trait that could appear in only one of two alternatives was passed from parent to child to grandchild *ad infinitum*.

At first in his lecture, Mendel concentrated on a single trait at a time. He discussed the odds of transmitting either the round or the angular pea shape. Then he talked about the odds of transmitting either yellow or green pea color, then seed coat tint (grayish or white), then height (tall or dwarf). He went on in this way, single trait by single trait, a total of seven times.

For each succeeding generation of monohybrid crosses, Mendel described how his thinking became more involved and why this compelled him to turn to more complicated mathematics. Here, no doubt, was where he started to lose his audience — and where he wisely decided to end the first lecture.

Mendel closed with a cliffhanger of sorts. He would return in March, he said, with an explanation of "the peculiar and regular way" in which these differentiating characters, which show up in the 3:1 ratio in the first generation of hybrids, separate out again in subsequent generations.

March 8, 1865, another Friday, was less frigid than February 8 had been, but in many ways the evening was a reprise. Mendel, again dressed in his frock coat and riding boots, lectured for an hour with numbers and calculations so percussive that some of his listeners thought their brains would pop. It was unclear whether he meant to draw conclusions about "heredity" — a word he never actually used. In the late nineteenth century, heredity was taken to mean constancy — the tendency of like to beget like, of pea plants always to create other pea plants, nightingales to hatch nightingales, blondes to give birth to blondes. Heredity was a synonym for stability, endurance, fidelity. But what Mendel was describing was not simply constancy. His laws applied not only to traits that were inherited intact from the parent plant, but to traits that differed from one generation to the next. He showed why not every puppy in the litter looked just like its mother, why when a blonde woman married a dark-haired man her children might not be blonde. For Mendel the laws of inheritance were not limited to descriptions of constancy; they described variability as well.

During the second talk Mendel lost the attention of most of his audience by focusing on some complex calculations — a form of "botanical mathematics" — that might have reminded them of the "mystical numbers" of the Pythagoreans. These were the followers of the fifth-century-B.C. mathematician Pythagoras of Samos, who is still remembered because of his theorem $A^2 + B^2 = C^2$, relating the lengths of the two sides and the hypotenuse of a right triangle. All things are numbers, said the Pythagoreans, whose favorites included perfect numbers, triangular numbers, and the number ten. Their favorite symbol was the pentagram, their favorite plant the bean, whose embryolike shape made them believe the beans were reincarnated babies. Indeed, Pythagoras saw many plants and animals as reincarnated humans. He was said to have stopped a man from beating a dog because he thought he recognized the dog as a reincarnated friend.

In the March lecture, Mendel went beyond the monohybrid crosses into the realm of the dihybrid and trihybrid crosses. It was here that the equations really got tricky. About three or four years into his experiment — that is, around 1860 or 1861 — Mendel designed crosses in which he considered two traits at once. After collecting the relevant data, he used algebra to calculate the number of possible combinations the crosses could result in and to derive the proportions in which these combinations were likely to appear. The math was simple, but the equations were long, boring, and often hard for his audience to keep track of.

He went back first to his original combination series, the one derived from the monohybrid cross *(Aa) × (Aa)*. The product of this cross, *A + 2Aa + a*, could then be multiplied by the product of a cross involving a different characteristic, represented as *B: B + 2Bb + b*. It was a matter of simple algebra, using the distributive property of multiplication, to come up with an expression to represent a dihybrid cross between plants that were hybrid for two traits. That expression, derived from *(A + 2Aa + a) × (B + 2Bb + b)*, was: $AB + Ab + aB + ab + 2ABb + 2aBb + 2AaB + 2Aab + 4AaBb$. Much later this would be simplified to the ratio 9:3:3:1, which became almost as familiar as Mendel's original 3:1. The ratio means that, based on appearance only, for every nine offspring showing dominant traits for both A and B, you get three that are dominant for A and recessive for b, three that are dominant for B and recessive for a, and one that is recessive for both a and b. Mendel never worked out this 9:3:3:1 ratio himself; one of his rediscoverers did, early in the twentieth century.

At this point in the lecture, Mendel described two central principles that would eventually come to be known as Mendel's laws. He was far too modest to call them laws, much less attach his own name to them. And it was unclear, from the way he presented his lectures, just how confident he was that they held true for most other plants.

First came his observation of the quality that was later called segregation. He began by outlining his idea that some factors in the germ cells, still unidentified, were able to pass on traits from parent to offspring. Whatever they were, he deduced, they separated while getting ready to pass to the next generation through the gametes of the parents. Mendel also deduced that the sex cells somehow changed from having a double dose of hereditary factors to having only one. He could not say exactly how such a thing occurred; no one could say until twenty-five years later, when the process of reduction division (meiosis) was described. All he knew was that such a change was necessary if his math was to work.

A related observation, which would come to be called the law of independent assortment, was that each factor that is passed from parent to offspring is passed alone, independent of any other factor.

At this second lecture, Mendel spoke briefly about his two years crossbreeding other plants besides *Pisum*, especially different species of beans *(Phaseolus)*, including crosses between the French bean *(P. vulgaris)* and the bush bean *(P. nanus)*. With some of these crosses, he was able to confirm the 1:2:1 ratio he had obtained for *Pisum*. But with others he could not.

Flower color was especially perplexing, he said. When he crossed the white bush bean with the red scarlet runner bean *(P. multiflorus)*, his results were not what he had expected. "Apart from the fact that from the union of a white and a purple-red coloring a whole series of colors results, from purple to pale violet and white," he said, "the circumstance is a striking one that among thirty-one flowering plants only one received the recessive character of the white color, while in *Pisum* this occurs on the average in every fourth plant." It was to explain such anomalies, in fact, that he presented his lectures in the first place, hoping to encourage some of his colleagues to take on similar experiments and to prove his findings either wrong or right.

At the end of the lecture — possibly sensing that some in his audience were losing the thread of his argument — Mendel offered some thoughts about speciation, still the most charged scientific topic on both sides of the English Channel. In doing so, he dipped his toe into the murky waters of Lamarckism, a subject of debate almost as hot as the debate over evolution. "No one will seriously maintain that in the open country the development of plants is ruled by other laws than in the garden bed," Mendel said. "Here, as there, changes of type must take place if the conditions of life be altered, and the species possesses the capacity of fitting itself to its new environment." But, he went on, by "fitting itself" he meant that the species undergoes a change in the gametes, through the units that pass on dominating and recessive traits in an even-handed, random matter.

If adaptive changes do occur in response to environmental influences, he said, they tend to be conserved, passed on to subsequent generations. And Mendel's experiments provided a theory that explained how this conservation occurs. "Nothing justifies the assumption that the tendency to form varieties increases so extraordinarily that the species speedily lose all stability, and their offspring diverge into an endless series of extremely variable forms," he said. To the contrary, the tendency is toward stability, with variation being the exception, not the rule.

In other words, Mendel found in his crossbreeding experiments the mechanism of species stability. And this, rather than the mechanism of variation, which drove the work of so many earlier hybridists, might be what he had been looking for all along.

Picture Gregor Mendel, now forty-four, intent on mailing out reprints of his first journal article. As was customary with lectures delivered at its monthly meetings, the Brünn Society published his complete paper in its official *Proceedings*. The paper's publication in 1866 gave him a second chance at scientific recognition, since news of the lecture had not spread beyond the city borders. So Mendel requested forty reprints from the journal editor, an excessive number at that time, and set about publicizing his results as best he could, within the limits of his naturally reticent personality.

His curly brown hair thinning around his widening face, Mendel sat at the oak writing table in the orangery, where the air was warm and lushly fragrant. Addressing envelopes in his methodical way, he gathered the nerve to send reprints to at least a dozen respected scientists throughout Europe. He may have distributed all forty, but we know for certain the fate of only twelve.

One went to Kerner von Marilaun, the botanist from Innsbruck who had attended Franz Unger's lectures on plant physiology at the University of Vienna at the same time Mendel did. In 1875 Kerner would begin a famous set of experiments. By transplanting

lowland plants to alpine habitats, he was able to prove that alti-
tude-related changes in the plants were not transmitted to offspring planted back in the lowlands. Kerner focused on highly
variable plants in his search for the source of speciation, and he apparently saw little relevance in the monk's work with the stable
garden pea. Nor did he show any interest in Mendel's career, although the two had crossed paths many times in Vienna. When a
copy of Mendel's reprint was recovered from Kerner's library after
his death, the pages were uncut — that is, the printed sheets had
been folded into page-sized segments but had not yet been slit
apart. Clearly the professor had never bothered to look at it.

Another uncut reprint was found in the library of Charles Darwin, so Mendel must have sent him a copy, too. But even if Darwin
had taken the time to cut through the folds and try to read Mendel's paper, he might not have understood it. Darwin had, after all,
been exposed to the work of Charles Naudin, who reached many
of the same conclusions Mendel had — although without the statistical proofs — and he had not been especially impressed. "He
cannot, I think, have reflected much on the subject," he once observed about Naudin. Why would he have been any more charita-
ble toward Mendel?

Indeed, Darwin himself obtained ratios similar to the Mendelian 3:1 ratio — but had no idea what they meant. In 1868, just
one year after Mendel's reprint arrived, Darwin published a paper
in which he mentioned the "prepotency" of a character trait. He
was breeding snapdragons at the time, crossing those bearing red
flowers with those bearing white. In the first hybrid generation he
found, like so many hybridists before him, that all the hybrids were
white. But in the second generation, he noted that a "minority"
of snapdragons revealed the red color of one of the grandparent
plants. Darwin, who was not in the least a quantitative experi-
menter, nonetheless counted the different colored flowers, noting

eighty-eight white plants to thirty-seven red — a ratio of 2.4:1. He did not calculate the ratio, but if he had he would not have known what to make of it without Mendel's insights. And there sat the monk's paper, uncut and presumably unread, on an out-of-the-way shelf.

A third reprint ended up in the private library of Martinus Beijerinck, a well-known Dutch biologist who later had a difficult time of his own over questions of priority. When Beijerinck announced in 1898 that he had discovered an infectious agent, which he called a virus, a Russian biologist, Dmitry Ivanovsky, quickly claimed prior credit for the discovery. Once Beijerinck was made aware of Ivanovsky's paper, published six years before his own, he conceded credit to the Russian. Like Mendel's, Ivanovsky's paper appeared in an obscure journal that never received wide readership and never was translated from its original language. Unlike Mendel, Ivanovsky was still alive when his findings were "rediscovered" by a more prominent scientist — and was feisty enough to insist on getting the credit he believed he was due.

Beijerinck, who received the Mendel reprint from an intermediary, eventually mailed it to a fellow Dutchman, Hugo De Vries. When Beijerinck learned that De Vries was about to publish a paper summarizing his work on hybridizing both *Oenothera lamarckiana* (evening primrose) and *Zea mays* (maize), he tracked down the reprint and sent it to his younger colleague. "I know that you are studying hybrids," he wrote, "so perhaps the enclosed reprint of the year 1865 by a certain Mendel which I happen to possess is still of some interest to you." This transmission took place some time between 1898 and 1900; the accounts vary. But the exact date is of some historical significance. De Vries, it would turn out, was taking the same path Mendel had, and in 1900 he was widely heralded as one of three "rediscoverers" of Mendel's paper. At that point the question of when he had encountered the monk's paper became relevant to the discussion of whether De

Vries had arrived at his own conclusions independently or by following Mendel's lead.

A fourth reprint arrived at the Max Planck Institute in Tübingen, Germany, by a circuitous route. It was sent originally to an unknown recipient, who sent it to Theodor Boveri, the co-developer of the chromosome theory of the cell. After his death in 1915, Boveri willed the reprint to the Kaiser Wilhelm Institute for Biology in Berlin. Coincidentally, the first director of the institute, Karl Correns, was another of Mendel's rediscoverers in 1900.

Did Mendel also send a reprint to Franz Unger, his botany professor in Vienna? A fifth reprint does exist in the library of the Institute of Botany at Graz University, where Unger taught before moving to Vienna in 1849. That reprint might have been donated by Unger, who might have received his copy from Mendel himself. By 1867, when Mendel was sending out reprints, Unger had retired from teaching. But there is no record that he sent Mendel a letter acknowledging the work he had done — nor that he even read the reprint, which, like so many others, was found uncut. Why would Unger turn his back on his former student? Wouldn't he have seen the genius of his experiments and want to help him get the results published in journals with a wider readership?

Apparently not. But whatever motivated Unger will never be known. From what he knew of Mendel's work, he might have considered it anti-Darwinian, with its emphasis on constancy over variability and its implied rejection of blending inheritance. If so, he would not have wanted to help disseminate findings that could be used to contradict a theory he had risked his own career to support. Or maybe, just three years before his death, Unger lacked the strength to engage in what was sure to be a demanding correspondence, both intellectually and emotionally, with a brilliant former student.

Perhaps reprint number six went to a botanist Mendel had first learned about through Unger's lectures. This man, M. J. Schleiden, was codiscoverer of the cell theory and the author of *Principles of Scientific Botany*, which in 1850 helped establish botany as a deductive science. He had boundless faith in the power of numbers, and he would have appreciated Mendel's methodology more than any other botanist alive at the time. In botany, Schleiden wrote, you cannot develop a complete theory of anything without mathematics.

Five more reprints have been found in the past forty years, but the paths that brought them from Point A — the St. Thomas monastery — to Point B — the library or collection in which they are now housed — are shadowy. One reprint is at the University of Indiana. Two more are in private collections in England, having been purchased at auction in the 1980s for 4,400 and 13,500 German marks, respectively. One is in the monastery library in Brno, and another at the National Institute of Genetics in Mishima, Japan.

The last reprint whose fate we know about ended up on the desk of Professor Karl von Nägeli at the University of Munich. Mendel sent the reprint on the very last day of 1866, along with a letter summarizing his eight years of *Pisum* experiments. "The presence of nonvariant intermediate forms, which occurred in each experiment, seems to deserve special attention," he observed dryly, in a letter no doubt typical of those he sent to other noted scientists to accompany his report. He explained his familiarity with the work of Gärtner and other earlier hybridizers, and shared results of his preliminary work with hybrids of *Hieracium* (hawkweed, one of Nägeli's favorite plants), *Cirsium* (prickly thistle), and *Geum* (a member of the rose family). "I lack [the right] kind of experience" for natural observation, Mendel wrote, "because the press of teaching duties prevents me from getting into the field frequently, and during the vacations it is too late for many things" — that is, the growing season of many wild plants would already have passed. For this reason, he asked the indulgence of better-educated, better-positioned men in providing assistance and advice as his experiments progressed.

After he had sent out the reprints, Mendel's work was over — at least until he could get back to his "children" when the growing season began again. All he could do was sit back and wait for the responses of his scientific correspondents. By this point, Brünn was again in the grip of winter, the winter of 1867.

Interlude

12

The Silence

A visitor to a garden sees the successes, usually.
The gardener remembers mistakes and losses, some
for a long time, and imagines the garden in a year,
and in an unimaginable future.

— "A Shape of Water," W. S. Merwin

MAIL CALL at the St. Thomas monastery was always an event. Every one of the dozen or so monks who made up the cloistered community was a learned man. Abbot Napp had seen to that, selecting the brethren as though forming a band of scholars at a mini-Chautauqua, in which each member was allowed the luxury of books, collegiality, and time to ruminate and dabble as the spirit moved him. In the 1860s, one of the best ways to feed this spirit was through written correspondence. Letters were the mechanism of discovery.

So twice a day, between New Year's Day of 1867, when he sent out his first letters, and the end of February, when he received his first reply, Gregor Mendel must have experienced a tiny flutter of anticipation, the slow dawning of disappointment, and finally the heavy thump of despair — only to go through the cycle again at that day's second mail call, or, after a long night of knowing that something was not quite right, at mail call the following morning. Nearly a hundred times the flutter and thump were repeated, becoming an almost familiar accompaniment to the passage of time through a season in which there were, finally, no new peas to count, no new puzzles to work out.

Of all the scientists to whom Mendel wrote in hopes of finding both help and validation, only Nägeli wrote back. And even he took almost two months to reply. With what excitement Mendel must have received that first letter on February 27, 1867. What visions must have danced in his head of a fruitful discourse with one of the most incisive minds in Europe. Nägeli, after all, had been the paragon of Mendel's esteemed Professor Unger. How flattering that he should take the time to write to poor lowly Mendel. Now it hardly mattered that Mendel had been checking the mail with growing trepidation every morning and every afternoon for weeks. Forgotten were those hundred tiny disappointments, piling up as Mendel waited for acknowledgment that his eight years of labor had not been in vain.

The letter was handwritten, of course; the typewriter had just been invented and would not go on the market for another seven years. Nägeli took great care in composing it, going through several drafts to get the wording right. He revealed some "mistrustful caution" about the priest's findings. He also enclosed reprints of five recent journal articles. Although his language was polite, it was skeptical — and even more skeptical were the earlier drafts, which are all that remain of what Nägeli actually wrote. In the notes he wrote to himself as he crafted his response, Nägeli wondered how Mendel knew that the hybrid he was calling Aa was a constant form. "I expect that (when inbred) they would sooner or later be found to vary once more," he wrote. "A, for instance, has half a in its body [being bred out of Aa], and when inbred cannot lose that element." These notes suggest that Nägeli misunderstood the core finding of Mendel's paper — that the recessive a determinant would of course remain in the hybrid and, rather than being lost, would eventually express itself, given the right combination of gametes, in subsequent generations.

In addition, Nägeli pointed out, Mendel's interpretation of his ratios might have gone beyond what the data allowed. "You should regard the numerical expressions as being only empirical," he wrote, "because they can not be proved rational." In other words, Mendel's ratios might have been sufficient to describe his experimental observations, but they were not solid enough to form the basis of any kind of general theory of inheritance, hybridization, or species formation.

What drove Nägeli's response? Was it malice, or shortsightedness, or vanity? He no doubt saw that if Mendel was right, then he, Nägeli, had to be wrong. Nägeli was a proponent of the idea that offspring get a bit of inherited information (which he called "idioplasm") from the mother and a bit from the father, and express the two in a form midway between them. He no doubt saw that Mendel's evidence could be interpreted as a disproof of blending. In Mendel's system the traits that seem to disappear — perhaps into a blended state, perhaps not — show themselves, in the next generation, not to have blended at all. If Mendel's data got out, what would become of Nägeli's considerable reputation?

Maybe Nägeli was just tired. He had always, even as a child in Switzerland, had a delicate constitution, and now he was nearly bedridden, suffering the long-term effects of the cholera he had caught ten years earlier in St. Petersburg, where he had gone to get ideas for organizing a new research institute at the University of Munich. His chronic illnesses might have been part of the reason for his tardy and churlish reply to Mendel's first letter, as well as his replies to several other letters later.

Nägeli's response may have chilled Mendel, who, for the short time between holding the sealed envelope in his hand and opening it, probably allowed himself to expect something more. But it just as easily may have encouraged him, not so much for the words themselves as for their provenance. The only way to gauge his reaction to Nägeli's first letter is by reading what he wrote in reply. In his second letter to Nägeli, written on April 18, 1867, Mendel described in greater detail his *Pisum* experiments, probably surmising that Nägeli had either not read or not understood the original paper. He clarified his belief that at least some of his hybrids would always breed true. "I have never observed gradual transitions between the parental traits or a progressive approach toward one of them," he wrote. "The course of development consists simply in this: that in each generation the two parental traits appear, separated and unchanged, and there is nothing to indicate that one of them has either inherited or taken over anything from the other." Finally, he defended his statistical findings with the blunt statement that it was "permissible" for him to move from observations to theory "because I have proved by previous experiments that the development of a pair of differing traits proceeds independently."

The correspondence between Mendel and Nägeli would sputter along for the next seven years — with long periods of silence, masking doubt and frustration, between one letter and the next. To Mendel's lengthy second letter Nägeli made no reply at all. But Mendel, undaunted, wrote the professor a third letter anyway, on November 6, 1867.

This time Mendel avoided writing about *Pisum* altogether, perhaps inferring from his silence that Nägeli was not especially interested in peas. Instead he focused on some of the species he knew Nägeli worked on himself, especially the hybrids of *Hieracium*. He also revealed a little bit more of his own gentle, self-mocking personality. In this letter he made fun of his "excess of avoirdupois" and admitted to an almost childish eagerness for summer. "[I am] impatiently awaiting [the season when] several fertile hybrids will exhibit their progeny in flower for the first time. Care has been taken that they may appear in large numbers, and I only hope that they will reward my anticipation with much information about their life history."

For all its charm, Mendel's third letter also met with silence.

What persistence, what self-confidence — or what desperation — it must have taken for Mendel to write again on February 9, 1868, three years and a day after he had first presented his paper. This time he took another tack. His letter was brief, asking only that Nägeli send seed or plant samples of fifteen species of *Hieracium* so Mendel could use them to produce whatever hybrids Nägeli might need. This was a wily move. Mendel was a highly skilled crossbreeder, and *Hieracium* was notoriously difficult to hybridize because its flowers were so slippery and small. Each flower has five stamens, which adhere to form a tube through which the pistil — the female structure — passes. The tube is so fragile that only the most steady-handed can successfully castrate the flower without also damaging the pistil. Because the hawkweed flower is so tiny, this painstaking work also requires a microscope, which can lead to debilitating backaches and eyestrain.

In effect, Mendel in his fourth letter offered his services as an unpaid research assistant — something no scientist, then as now, could easily refuse. The offer was especially welcome when it involved taking over a job as frustrating as the crossbreeding of *Hieracium*.

Mendel's proposal finally broke through Nägeli's silence. In April 1868 Nägeli's second letter arrived in the monastery mail. This one was not much more encouraging than the first, fourteen months earlier. It was brief and to the point, promising to send Mendel some plants and seeds as soon as possible.

A few weeks later, on May 4, 1868, Mendel wrote his fifth letter to Nägeli. But by then, everything had changed — and Mendel would never again have the time or energy to focus on the experimental work that had previously been his lifeline.

Five weeks earlier, on March 30, 1868, Mendel had been elected abbot of the St. Thomas monastery. Napp had died the previous winter at the age of seventy-five, and Mendel, in a close contest with his fellow monk Anselm Rambousek, was elected to succeed him. "From the very modest position of teacher of experimental physics I thus find myself moved into a sphere in which much appears strange to me," he wrote to Nägeli, "and it will take some time and effort before I feel at home in it."

While he managed to continue some experimental gardening, especially with the hawkweeds that Nägeli had sent, after the election Mendel's fate was basically sealed. For the last sixteen years of his life, he was an administrator first, a gardener only second. As for being recognized as a true scientist, that particular laurel continued to elude him. He would remain, as one observer phrased it, "a harmless putterer."

Mendel had loved the idea of becoming abbot, even though modesty and etiquette made him refrain from voting for himself. (He won on the second ballot, his own vote being the only one against him.) He was flattered to have been chosen by his brethren, calling his victory, in a rare unguarded moment, something for which he could "hardly venture to hope." But his election did not necessarily reflect any inherent leadership qualities or much confidence on the part of his fellow monks in Mendel's ability to shepherd St. Thomas through what were bound to be some difficult times. Other reasons no doubt came into play. For one thing, Mendel was the right age: a high property tax was levied on the monastery every time a new abbot was elected; men with many years to live were favored, so that changes in command would happen less often. (In 1868 Mendel was only forty-six years old; Napp had been elected abbot at the age of forty-two.) For another, he was of the right ethnic background. A struggle was raging between the German-speaking minority, which wanted to retain its hold on the reins of power through the workings of the Hapsburg monarchy, and the Czech-speaking majority, its nationalist movement in full flower. St. Thomas was traditionally a Germanic monastery, and Mendel was of German lineage. His opponent, Rambousek, was Czech.

Soon after being elected abbot, Mendel read Darwin's second book, *Variation of Animals and Plants Under Domestication*. The English edition was a tome of more than a thousand pages, bound in two volumes; just lifting it was a chore. In German translation it was smaller but no less impressive. Within days of Mendel's first setting his hands on it, the monastery copy of *Das Varieren der Tiere und Pflanzen im Zustande der Domestication* was littered with the pencil markings and scrawls that were the footprints of Mendel's intellectual meandering. Much the way scholars today do computer searches to check on how often their work has been cited, Mendel's marks often noted Darwin's references to one of his own associates or countrymen. But even though Darwin did extensive work with the garden pea, and spent four pages describing its habits and hybridization patterns, he never once mentioned the *Pisum* experiments of Gregor Mendel.

Darwin probably had never heard of Mendel. Nor had anyone but Nägeli — until 1881, when *Die Pflanzen-Mischlinge* by Wilhelm Olbers Focke appeared. The book, known familiarly as *Focke*, summarized the world's leading plant experiments and cited Mendel's work fifteen times. "Mendel believed that he found constant numerical proportions between the types of hybrids," Focke wrote, adding that the monk's work followed the tradition of early hybridizers. Like them, Mendel found that hybrids tend to revert to parental form and that old, apparently lost characteristics can reemerge generations later.

When George John Romanes prepared an entry on hybridism for the *Encyclopaedia Britannica* later in 1881, he borrowed a copy of *Focke* from his friend Charles Darwin. Romanes added Mendel's name to his list of plant hybridists, though he never read

Mendel's paper nor even Focke's brief description of it. Darwin never read Focke's description, either; in the book he lent to Romanes, the pages summarizing Mendel's work remained uncut.

Yet Mendel was not entirely ignored in his own lifetime. The problem was that some of the people who read and understood his paper were so obscure themselves that no one, including Mendel, had ever heard of them. I. F. Schmalhausen, for instance, was a graduate student in St. Petersburg who encountered Mendel's paper in the 1870s while working on his dissertation. Mendel's task, Schmalhausen wrote in a footnote to the German translation of his thesis, was "to estimate with mathematical accuracy the number of forms having originated from hybrid pollination and the quantitative ratio of the individuals of these forms." In carrying out this task, he wrote, Mendel "obtained complete series, the number of which can be represented as originating from a combination of several series [and he found that] constant members with new combinations of traits are always obtained."

Nothing ever came of Schmalhausen's recognition of Mendel's work — nor of Schmalhausen's own work, for that matter. The most significant citations were the fifteen in *Focke*, a widely read horticultural textbook, but those citations seem not to have been enough to encourage anyone to look up or try to understand Mendel's original publication.

In his sixth letter to Nägeli, written on June 12, 1868, Mendel finally felt comfortable enough to address the professor as "Highly Esteemed Friend" rather than "Highly Esteemed Sir." Perhaps his position as abbot had given him more self-confidence, more of a sense of himself as a peer even of university professors from Munich. Or perhaps he felt buoyed by Nägeli's long and collegial response, which came less than a week after Mendel had sent his fifth letter. In this third letter to Mendel, Nägeli proposed two draft classifications of the *Hieracium* hybrids he was studying and sent nineteen living hawkweed plants for Mendel's experimentation. From this point on, the correspondence between the priest and the professor centered entirely around *Hieracium* — which turned out to be a completely misguided choice of plant for Mendel to work on.

First of all, Mendel needed to work with lenses, mirrors, fine needles, and artificial light to see what he was doing. Second, he needed the most delicate touch to manipulate the plant. "Before the pollen ripens, the pistil and the stigma are extremely sensitive to pressure and injuries," Mendel said, "and even if they have not been damaged they usually wither and dry up very quickly as soon as they have been deprived of their protective wrapping." He visited his hawkweed plot each morning between seven and nine, hoping to catch the plants at just the right moment for cross-pollination. "Day by day, a new series of florets opens," he said, "the opening beginning at the margin and extending towards the middle of the composite." He made use of this pattern of opening by dusting the floret then with fresh pollen from another *Hieracium* species "as soon as the stigma showed itself."

But even more significant than the delicacy of the operation was a certain peculiarity of this plant genus. No one knew it at the time — although some people suspect that Nägeli might have, since he had studied *Hieracium* for decades — but the hawkweed usually reproduced in such an unconventional way that Mendel could not possibly have gotten meaningful results from it. Except in rare circumstances, *Hieracium* relies on an asexual method of reproduction known as apomixis. In animals, this is called parthenogenesis, from the Greek for "virgin birth." Instead of mixing traits from each of two parents, hawkweed offspring are exact replicas — clones — of the mother plant. So Mendel spent all those hours at the microscope, all that time hovering in the garden, nearly losing his sight in the process, in pursuit of a wild goose.

13

"My Time Will Come"

THE ABBOT COULD NOT really tell whether the problem was in his back, his shoulders, his eyes — or his head. All he knew for certain was that the hours he spent huddled over his microscope were no longer enjoyable. He had to drag himself to the orangery during May and June of 1869, lacking the raw enthusiasm that had propelled him there in other Mays and Junes. Was he perhaps coming down with something? Had he caught a chill on one of those wet early spring days when he had sown his *Hieracium* seeds? Were his abbatial responsibilities making him hesitate a shade too long before hastening to his scientific studies? Did he somehow feel guilty about being here in the orangery at all?

Mendel's sessions at the microscope were now punctuated by frequent breaks. He stood up often to stretch his back, rub his neck, and rest his eyes. The eyes were the worst: sometimes they buzzed so painfully that he could barely see.

He went on this way for weeks, peering into the brass microscope, standing up and stretching, trying to rest his eyes between sessions. The microscope was necessary for studying the *Hieracium* plants because the flowers could not be easily seen with the naked eye. But the microscope, ultimately, proved to be the source of the problem — and almost cost Mendel his sight.

"Since ordinary dispersed daylight was insufficient for my work on the tiny *Hieracium* flowers, I made use of an illuminating apparatus [a mirror and a lens], without thinking what damage I might thus do to my eyesight," Mendel wrote to Nägeli the following year, explaining the self-imposed six-month hiatus in his hybridizing. "Well, during May and June, when I had been busied upon *Hieracium auricula* and *praealtum*, I began to suffer from a peculiar sense of fatigue and tension in the eyes, and, though I now spared my sight as much as possible, this trouble became intensified, making me quite unfit to bear any sort of eye-strain until well on in the winter. Since then, I am glad to say, the trouble has almost completely passed away, so that I can now [this letter was written on July 3, 1870] read for long stretches once more, and have even been able to resume my fertilization experiments with *Hieracia*, in so far as these are possible without artificial illumination."

Eventually Mendel was able to resume his work on hawkweeds. But he might have been better off not bothering. Because of apomixis, the results Mendel got from *Hieracium* undermined his confidence in everything he had discovered before. He described his work to the Brünn Society for Natural Science in a lecture on June 9, 1869, and published it the following year in the same journal that had published his *Pisum* paper. But that was the end of it. Undone by the bizarre behavior of *Hieracium*, Mendel ordered no reprints of his second paper, sent no letters, sought no advice. If he could not reproduce his earlier results, his earlier results must have been wrong. Now Mendel doubted all his previous ideas about dominance, segregation, and the 3:1 ratio. Because of misdirection, either accidental or deliberate, from the great Nägeli, Mendel lost faith in his own results. And he never again had the courage to promote them in public — or, as far as we know, even in private conversations with his nephews or the other monks.

The Mendel-Nägeli correspondence continued for a few more years, but Mendel's heart was no longer in it. Not only was he ailing; he was busy. Far from freeing him for more hybridization research, his position as abbot had diverted a staggering amount of his attention. Mendel was responsible not only for running the monastery but for maintaining honorary positions with institutions throughout Brünn. The abbot of St. Thomas was automatically involved with the local school authority, banking authority, and scientific society — all of which took time. And, as a prominent citizen, the abbot was expected to do a good deal of entertaining. Every Sunday afternoon when the weather was fair, a group of local dignitaries came to the monastery to play skittles, a game of lawn bowling for which Mendel had had a special "skittle alley" built on the brewery side of the courtyard. From three o'clock until about seven, the lawns were filled with the chatter of some of the most influential men in town: Captain General Count Vetter von der Lilie; Councilors Januschka, Klimesch, and Ruber of the lord lieutenant's office; Dr. Scharrer, president of the supreme court of Moravia; Herr Schilda and Herr Strobach, councilors of the supreme court; Herr Pieta, manager of the state lottery; and Professor Rost of the Brünn Realschule.

Gardening was never far from Mendel's thoughts, however. For his abbatial shield, he chose images that linked his love of flowers with his love of knowledge and humanity. Like all such shields, his had four symbols, one in each quadrant. Mendel's had a lily to symbolize constancy; a plough and cross to symbolize both his own peasant origins and the Augustinian credo of charity; the Greek letters alpha and omega joined by an equals sign, for scientific inquiry; and two clasped hands, for friendship and community. The design soon became part of the *trompe-l'oeil* decoration of the library ceiling. At each corner of the ceiling, Mendel ordered the painting of clusters of different flowers, including roses, forget-me-nots, and — his favorite — fuchsias.

On October 13, 1870, two and a half years after Mendel became abbot, a freak tornado raged through Brünn. As he watched the wind whip across the courtyard, sending furniture flying, tearing apart trees and outbuildings, the meteorologist in him went into high gear. He paid attention, as any good scientist would, to every minor detail of nature's fury.

The air roared a "hellish symphony," Mendel wrote less than a month later, "accompanied by the crash of window-panes and slates, which in some cases were flung through shattered windows to the other side of the room." As he sat near the window in his study, he watched a piece of roof slate fly through his open door-way, hurtle across his desk, and pass into the next room. But despite the "terrifying" nature of the brief cyclone — Mendel calculated that it swept through his immediate vicinity for no more than four or five seconds — the abbot was able to make detailed observations, and to record them with his typical care.

"The storm consisted of two gigantic cones," he told the Brünn Society of Natural Sciences in a lecture on November 9, 1870, "the upper of which had its point directed downward, and it seemed to hang from an isolated roundish mass of clouds, not very large, a mass in which a marked unrest, a vigorous to-and-fro movement, was noticeable." He described the upper and lower cones, their shape, their coloration — and observed that, contrary to expectations, the tornado most definitely was spinning clockwise.

"This brings to an end the discussion of our dangerous guest," Mendel said in closing. "But we must admit that, however we might have tried, we have got no further than an airy hypothesis, which is explained from airy material and on an airy basis."

The abbot's light tone, typical of his informal discourses but missing from his earlier scientific lectures, indicates that perhaps he did not take this tornado account quite as seriously as he had taken his earlier lectures on crossbreeding and the laws of inheritance. Or maybe it indicates simply that Mendel was older and wiser, aware finally that no matter how straight and earnest you were, no matter how much you comported yourself according to the rules, sometimes things just didn't work out the way you expected them to, and to your contemporaries you turned out to be virtually invisible.

The biggest surprise was the precision, and the irony, of the storm's damage. Almost as though his two great loves, horticulture and meteorology, were waging war, the cyclone was most devastating to one particular structure on the monastery property: Mendel's greenhouse. Did it break his heart to see that the whirlwind — which many people in those days believed to be the work of the devil — turned his beloved greenhouse, to which he still held out the hope of returning, into a splintered ruin? Or did he realize that his idyll in the greenhouse was already over, that it had ended years before, when the steady thrum of activity that had breathed life into its humid air was stilled by circumstance and disappointment?

If Mendel was brokenhearted, he managed still to describe every nuance of the whirlwind with the same care he had used five years earlier to describe his peas. Both accounts relied on clear observation and an almost naive faith in the power of description to uncover the telling detail, the one that would somehow enable all the other details to fall into place. And in both accounts, not even the smallest gesture or the most perplexing incongruity was ascribed to the hand of God. This is remarkable, considering that these accounts were written by a priest, an abbot, the second-highest church representative, after the bishop, in the city. If there is a deity in either of these lectures, it is the deity of scientific observation, the presence of God through the power of deductive reasoning.

In 1874 all the monasteries in Brünn were assessed a new city tax. The St. Thomas monastery was ordered to pay the government a total of 36,680 guilders for the next five years. The very idea made Mendel boil. Casting aside his botanical work for good, the abbot began a single-handed letter-writing campaign against the monastery tax, which meant taking on the provincial government, the ministry of education in Vienna, and the Liberal government of Moravia — which he had helped vote into power, despite the Augustinian tradition of voting Conservative. His letters to the tax authorities grew more detailed, more impassioned, more strident and vituperative as the years went on. The dispute continued for nearly a decade — almost until the day Mendel died. The stubborn abbot never wavered in his insistence that a tax on church property was unconstitutional. His adversary, Baron von Possinger, the lord lieutenant of Moravia, believed Mendel's resistance was due to "a deplorable condition of mental tension." Mendel said he would "never dream of failing to comply with any instructions from the exalted lord lieutenancy" if he could convince himself that the orders were constitutional — but he could not. Both men were probably right.

It was as though Mendel were deliberately taking on a quixotic campaign against the church tax. Maybe he thought abbots from the other monasteries would join him in the struggle. Maybe he truly believed the tax was wrong and felt compelled to protest it even to his dying breath. Or maybe there was a deeper explanation. Here we have Gregor Mendel, a man from peasant stock, a man who never rises professionally above the level of substitute teacher, a man who fails every important test he takes, a man whose dreams of scientific acclaim are dashed again and again.

We take this insecure man and elect him abbot — an election he himself regarded with surprise that his "unimportant self" could achieve such status. Then we throw in all the trappings of the position: a directorship at the Moravian Mortgage Bank, membership in the Royal and Imperial Order of Francis Joseph, curatorship of the Moravian Institute for Deaf Mutes, invitations to dine with some of the city's most prominent citizens.

Remember, this is a man so shy that in almost every official monastery photograph he is holding a flower, usually a fuchsia, as a sort of talisman — or perhaps as a way of placing himself, at least in a photograph, above the rank of the ordinary. This is a man so opposed to pomp that he never sat for an official portrait; the portrait hanging today in his former monastery is a reconstruction from a few different photos. And remember, this is a man who so disliked ceremony that his favorite activity was kneeling in the earth and getting his fingernails dirty.

Give him the rank of abbot, then; make him sit on all these boards of directors and official committees; have him entertain important men in the monastery garden on Sunday afternoons. A few years of this could easily have proved to be too much for a man like Mendel. How tempting to take on a cause that would all but guarantee that the important positions and pompous men would gradually slip away.

In the end they did, and Mendel was left with only a few loyal visitors. Other than his manservant, Josef, and his housekeeper, Frau Doupovec, who was with him when he died, Mendel became suspicious of everyone, even his fellow monks, whom he thought to be "nothing but enemies, traitors and intriguers." In the end he could count on only three young men to stand by him: his nephews, the sons of his beloved sister Theresia.

Johann, Alois, and Ferdinand Schindler attended the Gymnasium in Brünn at Mendel's expense and lived in the Klosterplatz, just opposite the monastery. They often spent weekday afternoons with their uncle Gregor, in the abbot's quarters or the garden, and on Sundays they joined the skittles games or sat with their uncle looking at pictures, chatting, or playing chess. The youngest, Ferdinand, was an especially challenging chess opponent, who went on to compose and publish many chess problems of his own.

In his final years only his nephews came. Mendel had made too many enemies — some of them his own guests — with his stubborn resistance to the monastery tax. Was it a relief for him to no longer have to maintain a lively public persona? "We cannot say Uncle was a misanthrope," Alois Schindler, the middle nephew, recalled after Mendel's death. "Only in his last years, after many disappointments which he lived through, he became diffident and stood aloof from society." An indication of what this solitary fight cost him is the change he made, in the closing years of his life, in one of the symbols on his abbatial shield. Borrowing a symbol from the shield of his predecessor, Abbot Napp, he replaced the clasped hands of friendship with an arm holding a cross, a traditional representation of piety. After a lonely, bitter struggle, Mendel apparently no longer felt part of a community of brethren, colleagues, or friends.

Even at the end of his life, even daunted as he was by personal and professional disappointments, Mendel maintained a delicious, somewhat mischievous sense of humor. He collected good jokes the way Darwin collected barnacles, underlining the best ones in the humor journal *Die Fliegende Blätter* ("The Flying Leaves") to read to his brethren at the dining table.

A sharp sense of humor is often taken to be a mark of genius. Great thinkers have talked about the fluid boundaries between discovery and comedy; the jester has long been seen to be the brother of the sage. The words "wit" and "witticism" have the same Old English root, *wītan*, meaning "ingenuity" or "inventiveness." In German, as in English, a single word, *Witz*, means both intelligence and humor; *Wissenschaft* (science) is close to *Aberwitz* (cheekiness), and both derive from the word *wissen,* to know.

Mendel was widely known for his wit. He loved to play jokes on his fellow monks, such as one monk named Clemens, who would often accompany the abbot on his strolls up and down the monastery grounds. One early March day, with the garden still under snow but the sun promising spring, Clemens and Mendel stood at the beehives of which Mendel was so fond. Dozens of his "dear little animals," as he called his bees, were already venturing from the hive where they had spent the winter, seduced outside by the warm sun. Mendel knew the habits of the bees; Clemens, unfortunately, did not. Lay your biretta down in front of the hives, Mendel said, smiling roguishly. As the two men looked at the round black hat lying stark against the white snow, the stunned young cleric saw his biretta turn from black to yellow. "The bees," Clemens recalled years later, perhaps retrospectively amused by the abbot's prank, "used it as a site on which to void that which, for reasons of cleanliness, they had refrained from voiding in the hives during the winter."

Usually Mendel kept his impishness in check. His more typical manner was one of friendly reserve, a veneer of warmth tempered by an underlying privacy. This was how he greeted strangers as well as friends. In 1878, for instance, he was paid a surprise visit by a young seed salesman from France. Mendel received him courteously, and gave his visitor a full tour of the gardens and an invitation to lunch. But there were certain subjects he simply would not discuss.

Look at this beautiful bed of green peas, said the young French salesman, C. W. Eichling of Nancy. How do you manage to breed them so they are so productive?

"It is just a little trick," Mendel said dismissively. "But there is a long story connected with it which it would be too long to tell."

Mendel began to explain a little bit — about the twenty-five varieties of peas he had imported, about disappointment with their yield because so many were dwarf bush-type peas rather than tall plants that bore more prolifically, about crossing these bush types with tall local sugar-pod types. But suddenly he stopped himself and awkwardly changed the subject.

Mendel was "one of the best-beloved clerics in Brno," Eichling recalled more than sixty years later, when he himself was a very old man and the abbot had achieved the international acclaim that eluded him in his own lifetime. "But not a soul believed his experiments were anything more than a pastime, and his theories anything more than the maunderings of a harmless putterer."

Although Mendel was forced to give up serious scientific gardening after becoming abbot — a decision that pained him for the rest of his days — he continued to dabble in ornamental horticulture. One of his favorite flowers to crossbreed was the fuchsia, the flower he holds in so many of his photographs. A local plant breeder, J. N. Twrdy, named a variety after him. The 'Prelate Mendel' cultivar, from the species *Fuchsia monstrosa*, was, according to the seed catalogue, a huge, luxuriant, early-blooming, unusually beautiful flower with pale blue petals "shading into violet."

Mendel was also forever amusing himself with scientific and mathematical ideas that had nothing to do with plants. On the back of a draft of one of his dozens of church-tax missives, he scribbled lists that show that, even in the midst of administrative tasks, he set himself new intellectual challenges. One of the most intriguing was a list of common surnames. Using several directories — the military yearbook of 1877, the register of transporters, the register of bankers, a barristers' yearbook — Mendel collected more than seven hundred names, which he arranged in different ways in an apparent attempt to spot some sort of pattern. First he placed them in alphabetical order, then he grouped them according to meaning. He thought of the names as nonbotanical hybrids, with different words attached to three common suffixes: *mann* ("man"), *baucr* ("farmer"), and *mayer* (a variant of the word for "butcher," in common usage meaning "fellow" or "guy").

The *mann* hybrids took up most of his list. He grouped nearly three hundred such names according to categories: craftsmen (builder or carpenter — *Baumann, Zimmermann*), officials (bureaucrat or customs officer — *Amtmann, Zollmann*), medical men (physician or healer — *Arztmann, Heilmann*). He also grouped names that recognized wealth (gold or treasure — *Goldmann, Schatzmann*), height (tall or high — *Langmann, Hochmann*), body characteristics (breast or neck — *Brustmann, Kehlmann*), or personality (laughter or happiness — *Lachmann, Frohmann*).

Maybe Mendel was idly passing time in the library study room, the orangery, or, now that he was abbot, one of the half-dozen or so rooms at his disposal in the separate prelate's wing of the monastery. Maybe he was hoping to entertain a dinner party with his organizational speculations. Or maybe he was toying with yet another variation on his conviction, stimulated more than twenty-five years earlier, that the numerical relationships of combination theory could be applied to any set of components in the natural or manufactured world.

At two o'clock in the morning on Sunday, January 6, 1884, Gregor Mendel died. His housekeeper, Frau Doupovec, was with him at the end. He had suffered for months from edema brought on by Bright's disease, a disease of the kidneys, and his legs had to be constantly wrapped in bandages, which soon soaked through with the excess fluid his kidneys were unable to excrete. "Your Grace, today you have already no water," Frau Doupovec told the abbot in the wee hours of the night. "Yes, it is already better," he replied. When next she looked around, Mendel was dead. He was sixty-three years old.

The obituaries were respectful. Few made reference to the tax struggle that embittered his final years; instead, they focused on his gardening, his weather watching, and his beekeeping. They sounded much like obituaries written for any high-ranking church official. "His death deprives the poor of a benefactor, and mankind at large of a man of the noblest character," read the notice in the *Tagesbote*, "one who was a warm friend, a promoter of the natural sciences, and an exemplary priest." Only in passing did anyone mention his experimental work. One eulogy, offered by Gustav von Niessl to the Brünn Society for Natural Sciences, whose January meeting was by coincidence held the very evening Mendel died, made reference — without much apparent understanding — to the abbot's "independent and special manner of reasoning."

Niessl failed to say then something he reported many years later, after Mendel was heralded as the father of genetics. During his years of anonymity, Niessl told Mendel's biographer, the priest was fond of saying to his friends, *"Meine Zeit wird schon kommen"* — "My time will come."

Shortly after Mendel's death, all his personal and scientific papers were burned in a huge bonfire in the monastery courtyard on the very spot where his greenhouse had once stood. The book burning might have been kindled by the jealousy of Mendel's successor as abbot, Anselm Rambousek, who had always disliked him, especially after Mendel edged him out of his first attempt to be elected abbot back in 1868. And, in a final twist, Rambousek, by showing a willingness to play along with the authorities that Mendel could never quite muster, ended the monastery tax struggle in a way that was more beneficial to St. Thomas than his predecessor could ever have imagined. With a few letters, Rambousek convinced the government that not only should the monastery be exempt from contributions to the loathsome "religious fund," but that the government actually owed the monastery for payment of back taxes in the amount of 19,876 florins. They sent him a check.

The burning of Mendel's papers might also have been nothing

more than routine housekeeping. Neither of Mendel's two surviving nephews had come to claim his papers, although Alois later said he had been waiting for Rambousek to offer the writings Uncle Gregor had once promised him. It was perhaps to be expected that the new abbot would want to clear away the papers of his predecessor, in which no one seemed especially interested, to make way for the ample accumulation of his own.

Today one of the few clear markings of Mendel's earthly passage is his grave in the city cemetery, a five-minute tram ride from Mendel Square on the far side of the River Svratka. The cemetery sits at the intersection of two six-lane roads, and the roar of traffic is audible even within its gates. Visitors shuffle among the gravestones, bearing flowers.

In the northeast corner, where traffic sounds are especially loud, a large plot is reserved for the graves of members of the Augustinian monastery of St. Thomas. Nine priests are buried here, their stones worn down by time and weather. A big marble monument stands at the center of the plot; a big iron gate surrounds it. Engraved on the monument is a line from the book of Romans: *Sive vivimus, sive morimur, domini sumus* — "Whether we live or die, we are the Lord's." And there, off to the far right, the oldest of the grave markers is Mendel's, its nearly illegible letters smudgy with mold. Not exactly the kind of stone one would expect to mark the burial place of the man who inspired one of the most profound sciences of our time.

Mendel might be embarrassed to see himself turned from a quiet, obscure, and brilliant man into the larger-than-life heroic figure he has become today. The story of this transformation, which erupted in a few great confrontations between 1900 and 1906, would change the shape of biology for the next hundred years. And it ended in irony: by recasting Mendel as an inspired genius unappreciated by his peers, his twentieth-century disciples undercut the very thing that marked his true contribution. Mendel was a dogged worker, not a hero — and it was his nonheroism that allowed him to do the plodding, patient, thorough work through which his genius emerged.

But maybe Mendel would not have been embarrassed. He was human, after all, and not completely free of vanity; remember that he sent out reprints of his *Pisum* paper, hoping that his accomplishment would be recognized. So maybe he would be pleased to find that the science named in his honor, Mendelian genetics, is at the heart of a revolution not only in biological thought but in thought itself. How close his final reward seems to the lines he used as a schoolboy to describe Gutenberg's reward: "That of seeing, when I arise from the tomb, / My art thriving peacefully / Among those who are to come after me." How unexpected, how glorious, his life-after-death would seem to a man who was haunted by a silence that echoed ominously for nearly twenty years.

Because for Gregor Mendel there would be, as there is only rarely on this earth, a second act.

Act Two

14 ✺

Synchronicity

And then the day came when the risk to remain
tight in a bud was more painful than the risk it took
to blossom.

— Anaïs Nin, 1903–1977

THE EVENING PRIMROSE is a messy little shrub. Its flowers are smallish and round, either yellow, white, or pink, possessing no special flair and emitting no special scent. But during the time of its blossoming, as midsummer's eve approaches and recedes, it sparks the air at sunset with a special kind of magic.

For most of June and July, each night as dusk descends, the evening primrose bursts into bloom. It happens all at once, in an uncanny synchronous movement that raises questions about the effects of time, light, and surroundings even on something as apparently insensate as a flower. Every night brings a display of floral fireworks. Each flower changes in an instant, one after another after another. Pop goes one flower; pop goes the next at almost the same moment, bursting open so quickly it seems to be spring-loaded. The petals start out folded like a loosely wrapped cigar; with one pop they form a pinwheel; with the next, an open four-petal cup. Each new bloom lives the night and into the next morning, and then it shrivels and dies.

Demeter, the Greek goddess of the earth, was said to be responsible for the evening primrose's nightly blooms. "The goddess

Demeter tended the countryside like a garden," goes the myth. Every spring she planted seeds, watered the earth, and urged trees to blossom and bear fruit. As Demeter worked, her beautiful daughter, Persephone, played nearby, picking flowers and singing in the breeze. Then dusk descended, and "when mother and child walked home hand-in-hand at the end of another sunny day, talking and singing and laughing together, the evening primroses opened just to watch them pass by."

The evening primrose, with its mysterious innate sense of timing, attracted the attention of some of the most original botanists of the late nineteenth century. One, Hugo De Vries of Amsterdam, became a special expert in this flower, whose formal name, *Oenothera lamarckiana*, honors the now-discredited biologist who believed that acquired traits explained evolution. But sometime before 1900, when De Vries had turned his attention from *Oenothera* to *Zea mays*, he stumbled across an old journal article that seemed to have anticipated his own conclusions by thirty-five years. The way De Vries handled that surprise, and the way two of his colleagues did when they chanced upon the same article at almost exactly the same time, is what gave Gregor Mendel a second chance. But it also kept De Vries from being recognized as the developer of the mutation theory and as a brilliant researcher in his own right. Now we remember him primarily for his role in the startling spring of 1900, when a new century began and when three biologists reached a single conclusion with as much uncanny synchrony as the opening of blossoms.

15

Mendel Redux

A garden, like a life, is composed of moments.
I wish mine could always be as it is right now, this
late afternoon at the end of March.

— *A Full Life in a Small Place*, Janice Emily Bowers

KARL CORRENS WOULD LATER remember the moment when he received the reprint by his Dutch nemesis as one of gut-wrenching anger — though he never expressed it quite that way. With Bavarian stolidity, he would merely say that for the previous six months he had been working on a paper describing the "complicated relationships" among hybrids in maize plants and that it was a great surprise to find himself preempted by the botanist from Amsterdam. Hugo De Vries's article described the same hybrid relationships in more than a dozen species, including the evening primrose, poppies with either black or white petals, and, in passing, the same species on which Correns was doing most of his work, *Zea mays*. "That other investigators also worked in the same direction I naturally did not know," he recalled twenty-five years later, in a comment that must have echoed Charles Darwin's feelings when Alfred Russel Wallace sent him his essay on natural selection; "otherwise I would have hastened more with the preparation of the publication."

This understatement masks the frenzy with which Correns read the journal article that arrived in the morning mail on April 21,

1900, a Saturday. Spring was just starting to dab bits of green along the trees and grasses in Correns's garden in Tübingen. At thirty-five, he was a lecturer at the university, having studied with some of the leading botanists of his time, including Karl von Nägeli at the University of Munich. Decades later the ghost of Nägeli still influenced Correns's life — not only because Correns would soon collect and publish the correspondence between his old professor and Gregor Mendel, but because his wife of eight years, Elizabeth Widmer, was Nägeli's niece.

Correns was a creative and hard-working researcher who had already achieved some international acclaim. But in his unguarded moments he worried that his career was stalled. Self-doubt took over that April morning when he read the short article by De Vries, published the previous month in *Comptes Rendus de L'Academie des Sciences*, the official journal of the French Academy of Sciences. The paper had been read aloud to the Academy by G. Bonnier (whose French accent was almost assuredly better than De Vries's) on Monday, March 26. Correns read the two-page paper in one breathless gulp. He was furious that De Vries, of all people, had beaten him in the race for publication. The two men had been in similar races before, and De Vries had always won. Correns's fury grew as he realized that De Vries did not understand the significance of the ratios he had derived. And then the last straw: De Vries did not even credit the scientist who had gotten there before either one of them — Gregor Mendel.

In Correns's opinion De Vries had overlooked the essential elements of inheritance. He echoed some of the best points made by Mendel, such as the tracking of individual characters rather than species characters in successive hybrid generations. He even went beyond Mendel to some of his own innovative concepts, strongly setting forth an idea we still are not sure that Mendel held: that character traits are transmitted by material elements. But he conveyed no apparent understanding of the significance of Mendel's

3:1 ratio and no apparent interest in determining how often this ratio appeared or whether it constituted the essential rule of inheritance — or was merely an exception to the rule.

How infuriating for Correns to be beaten to publication by a man who did not even get it. How infuriating, indeed, to be beaten again by De Vries, who had been first to publish about a different topic only the year before. In 1899 both men were working on what was known as the "xenia problem." Xenia was a puzzling phenomenon involving a plant's endosperm, the placentalike cell layer that nourishes the growing embryo. Endosperm was thought to arise from exclusively maternal tissue, yet when pollen from another plant was introduced during cross-fertilization, the endosperm seemed to change. This effect of foreign pollen on a plant's endosperm was known as xenia. It was a puzzle: how could pollen have any effect on something that theoretically belonged only to the female portion of the plant?

In the late 1890s xenia was one of the most vigorously researched questions in botany, the subject of seven major publications or lectures in 1899 alone. Interest in xenia was sparked that year by two independent papers presented at scientific meetings in August 1898 and March 1899, the first by a Russian and the second by a Frenchman. Both proposed a theory that flowering plants undergo "double fertilization" — the first time to create the embryo, the second to create the endosperm. These two papers produced what was later called "the botanical sensation of 1899."

De Vries and Correns both joined the race to publish papers about xenia and double fertilization. And De Vries beat Correns to the finish line. Now De Vries seemed to have beaten him again. This time Correns planned to do something about it.

The events of the spring of 1900, when Mendel's thirty-five-year-old paper was "rediscovered," played out with almost operatic inevitability. Three men, three countries, three lines of thinking, and, after traversing their independent pathways, each arrived at the same place at virtually the same time. Of course their names would be tethered in the history books — to one another and to Mendel — from that moment on. The coincidence was too juicy to overlook.

But tales of the dramatic rediscovery are marred by uncertainties. When exactly did De Vries read Mendel's paper? Was it after his experiments had led him to a basic understanding of traits and determinants, as he said? Or was it before, just as he was designing his own research (and, if Correns's innuendo is to be believed, possibly copying Mendel)? And what about Correns? Did his teacher Nägeli really fail in the 1880s to tell him anything about Mendel, even though Nägeli and Mendel had corresponded for seven years, even though Correns's research model was *Pisum sativum*? And the third and youngest rediscoverer, Erich von Tschermak — why did he work so hard to get himself recognized as a "rediscoverer"? Did he even understand, in the paper he published in June 1900, how his findings stacked up against Mendel's ideas? Or was he primarily thinking that his best shot at immortality was to claw his way into the rediscoverers' circle?

These questions are not just about historical arcana. To learn how knowledge advances, we must understand how scientists use their predecessors' results. "If I have seen further," Isaac Newton supposedly said, "it is by standing upon the shoulders of giants." So it is with most great scientists. They can see things their predecessors could not precisely because of what their predecessors *did* see. Those who follow try to wrest the truth from the findings of those who came before. Afterward they disseminate new truths, hoping someone else will stand upon *their* shoulders and carry on. Usually, when scientific progress works this way, it is not called rediscovery. It is called, simply, discovery.

The story of the spring of 1900, though, unfolded a bit differently. The shoulder analogy begins to fall apart, and what we are left with is a scattering of isolated scientists each pursuing a similar goal. Mendel stood for a while on some giant shoulders, like Gärtner's and Kölreuter's, but the genius of his particular approach — applying the quantitative methods of math and statistics to his botanical findings — had no antecedent, and he was left to draw conclusions completely on his own. And Mendel's own giant shoulders offered no support to the men who came after him, because his concepts were so far ahead of their time — and because the very existence of Mendel's research results was practically unknown.

The rediscoverers quickly made amends for this oversight. The confluence of events of 1900 — loud debates about evolution, the race to understand inheritance, rancorous priority claims — made Mendel's followers eager to wrap him in the cloak of posthumous greatness. Within the year he was enshrined as an unappreciated genius born too soon, the guiding spirit behind the nascent (and still unnamed) science of genetics. And those who came in his wake had to accept the apostolic roles of rediscoverers rather than discoverers in their own right. This was a role De Vries chafed against, a role Tschermak longed for, and a role Correns virtually created, locked as he was in a struggle having nothing to do with Mendel and seeking — for both himself and his competitors — a dignified way out.

In 1900 Hugo De Vries was fifty-two years old and the most respected botanist in Holland. He was a professor at the University of Amsterdam and director of its Botanical Institute, located next door to the city's sprawling Hortus Botanicus (Botanical Gardens) along the Nieuwe Heerengracht canal. De Vries's associates respected and feared him but did not especially like him. They spread stories about his discomfort around women, claiming that when he was alone in the lab at night, he would spit in the culture plates of his female assistant, hoping to throw off her results.

Whether the rumor was true or not, it indicates how distant his staff felt from the grand, dapper man with the untamed beard.

One of De Vries's few defenders was his heir apparent, Theo J. Stomps. Stomps was fatherless, and De Vries, with no children of his own, took care of the younger man as if he were a son, paying for his education and assuring him a lifetime job. The rumor mill worked overtime regarding Stomps; the story was that De Vries, a closeted homosexual, was in love with him, and promoted him not because he was deserving — he was neither bright nor especially hard-working — but because he was so handsome. If De Vries really was gay, this fact might have fueled his dedication to his mutation theory; living in a time that defined homosexuality as a disease, he might have seen his sexual impulses as a kind of mutation. There is no evidence, though, that he ever acted on those impulses in the repressive atmosphere of Holland in the early 1900s.

Whatever his personal story, no one doubted that De Vries was a brilliant botanist. In 1889 he published a theory of inheritance he called "intracellular pangenesis" — a deliberate recapitulation and improvement of Darwin's inadequate ideas about pangenesis. He emphasized individual characters, as Mendel had, rather than species characters, as did most other botanists of the day, and said they were passed on by material units, which he called "pangens," a refinement of Darwin's "gemmules," in honor of the great British naturalist.

Intracellular pangenesis, an inspired theory, has been borne out in many of its particulars by the modern understanding of genetics. In his later writings De Vries would promote the idea that a particular pangen accounted for a particular trait, no matter what species it appeared in. This notion — that the "hairy" pangen, for instance, works the same way in thistles and in greyhounds — underlies much of today's research using plant or animal models to draw conclusions about human health and disease. De Vries supported his ideas through his work with *Lychnis* (campion), in which he introduced the trait for hairlessness from one species *(L. vespertina glabra)* into a normally hairy species *(L. diurna)*. His ability to do so, he said, confirmed "the principal thesis of pangenesis, that the same hereditary qualities in different species are tied to the same material elements."

But for all his inspiration, De Vries had certain blind spots. In particular, he refused to believe that anything of much hereditary significance took place inside the cell nucleus — where, it turns out, the genes and chromosomes reside. He conceded that the pangens were made in the nucleus but said that most of their movements and activities occurred in the nonnuclear parts of the cell.

In the late 1890s De Vries began work on an even bigger project: his theory of mutation. He believed that mutations — or, as he called them, "monstrosities" — occurred at random and for unknown reasons and that their existence was the driving force of evolution. He had started working with monstrosities to confirm pangenesis, showing that discrete units must be responsible for the monstrosities he found in *Lychnis, Linaria,* and other species. But in the late 1890s he stumbled upon a group of wild plants bursting with monstrosities in a field just outside Amsterdam: the plants were evening primrose, *Oenothera,* in particular *O. lamarckiana.* With that species De Vries amassed evidence about the relationship between mutations and the creation of new species.

He planned a series of publications on the subject, building the drama as he worked toward a crescendo timed to coincide with the arrival of the twentieth century. At that point, 1900 exactly, he expected the first of his two volumes titled *Die Mutationstheorie* to appear.

In keeping with his plan, De Vries went to London in July 1899 to present his findings, titled "Hybridizing Monstrosities," to the Royal Horticultural Society, which was hosting its first International Conference on Hybridization and Plant Breeding. He reported his early findings from crosses of *Oenothera*, opium poppies *(Papaver somniferum)*, and *Lychnis*. The presentation went well, but his conclusions were sometimes confusing, and not only because his English was so heavily accented. Take his analysis of his *Lychnis* crosses. When he had crossed the hairy species *(L. diurna)* with the smooth one *(L. vespertina glabra)* back in the early 1890s, he said, he had gotten the expected F1 generation of hairy hybrids. The following year, after these plants self-fertilized, De Vries said that of the 153 F2's, 99 were hairy and 54 smooth.

Two years before the London meeting, he had reported the F2 ratio accurately: two-thirds hairy, one-third smooth. But in 1899 he claimed that those same numbers, 99 to 54, represented a 3:1 ratio — even though they were much closer to a ratio of 2:1. Had he read Mendel's paper in the interim and then reinterpreted his own conclusions?

In any event, De Vries admitted to his London audience that much research still needed to be done. "Very little is known," he said, "of the way in which such a transferring of characters takes place."

At the same meeting was William Bateson, a lecturer in zoology at Cambridge University. At the time Bateson was one of England's leading defenders of the so-called "discontinuous variation" school of thought regarding the raw material for evolutionary change. He was embroiled in a heated controversy with a group of scientists, including Francis Galton, who said that Darwin himself thought variation was "continuous," that it occurred in small increments, and that these slight, almost imperceptible adaptations were the raw material on which natural selection exerted its selective pressure in the struggle for survival.

Bateson, on the other hand, insisted that evolution proceeds by fits and starts, with nothing smooth or gradual about it. He devoted himself to finding proof of variations that were "discontinuous," that moved from oddball parent to oddball offspring at a pace resembling not a glacier but an avalanche, not a stately canter but a gallop.

Bateson took to De Vries immediately, seeing his monstrosities as proof of discontinuous variation. De Vries was an "enthusiastic discontinuitarian," he wrote to his wife, who was away when the Dutch botanist came to stay at the Bateson home in Cambridge en route to the London meeting. In Bateson's world view, his characterization of De Vries was shorthand for many things: smart, sophisticated, well-bred, one of "us." Anyone who was not a discontinuitarian had to be one of "them." And Bateson, an elitist to his bones, had a great, abiding disdain for "them."

For his part, De Vries could not have failed to be impressed with the power of the paper Bateson delivered before the assembled horticulturists on Tuesday, July 11, 1899. Two questions are germane in trying to explain how new species arise from hybridization, Bateson said: What is the mechanism by which new forms come into being? And by what mechanism do they persist rather than regress to their original forms or to some form midway between the original and the new?

"At this time," Bateson announced, "we need no more *general* ideas about evolution. We need *particular* knowledge of the evolution of *particular* forms. What we first require is to know what happens when a variety is crossed with its *nearest allies*. If the result is to have scientific value, it is almost absolutely necessary that the offspring of such crossing should then be examined *statistically.*"

All the italics are Bateson's — probably his own effort to imitate, in the printed version of his spoken remarks, the forcefulness that came to him so naturally during his heartfelt orations. The italicized words — especially *statistically* — are worthy of this sort of emphasis; strung together, they read almost as though Bateson were already familiar with Mendel's paper and were simply giving a summary.

The prefiguring grew even more uncanny as Bateson went on. "It must be recorded how many of the offspring resembled each parent," he said, "and how many shewed characters intermediate between those of the parents. If the parents differ in several characters, the offspring must be examined statistically, and marshaled, as it is called, in respect of each of those characters separately. Even very rough statistics may be of value."

The day after De Vries and Bateson presented their papers, a botanist named R. A. Rolfe lectured on the history of hybridization from the viewpoint of systematic botany. He made reference to Mendel, whose name had been unspoken in scientific circles since his death fifteen years before. Referring to *Hieracium* hybrids, Rolfe said in passing that "G. Mendel raised several artificially." This was the first recorded oral reference to Mendel before the rediscovery. We can assume Bateson and De Vries were in the lecture hall that afternoon, but if they heard Rolfe's mention of Mendel — along with the names of a series of other hybridizers — they apparently did not quite register it.

As Bateson and De Vries departed from London, each man probably decided to keep a close watch on the other. They both carried themselves with a haughty arrogance — in both cases masking, as such a stance often does, an underlying insecurity — that made them too similar ever to become real friends. But Bateson and De Vries were, at least in 1899, intellectual soulmates. For each man, mutations held the key to unlocking some of biology's biggest perplexities.

Just a day after Correns received De Vries's reprint from *Comptes Rendus*, he had composed a written response. By Sunday evening, April 22, 1900, he had already mailed it to the most prestigious botanical journal in Germany, *Berichte der deutschen botanischen Gesellschaft.* The following Friday, April 27, Correns delivered the paper to a meeting of the German Botanical Society. By then the *Berichte* had published a German version of De Vries's paper, in which he finally credited Mendel — though in a churlish footnote, thrown in almost as an afterthought. Scholars are still analyzing the footnote to see whether he had always meant to give Mendel credit or did so only at the last minute, when he heard about Correns's rage. Yes, he was using Mendel's terminology, De Vries admitted. But exposure to Mendel had not been necessary for the deductive process outlined in his report. The monk's "important treatise is so seldom cited," read the footnote, "that I first learned of its existence after I had completed the majority of my experiments and had deduced from them the statements communicated in the text." This statement was meant to short-circuit any suggestion that De Vries had merely replicated Mendel's experiments — rather than having arrived at the same general laws of hybridization through his own independent pathway.

For Correns's part, the fury that drove him during those two days of writing his response was a rare thing for him. He was almost always gentlemanly in his demeanor. As he gazes out of a photograph from 1905, everything about him seems placid, even meek: his wire-rimmed spectacles, the same shape and color as Mendel's; his soft lips; the unruly hair combed carefully across his high forehead; his oddly sloping nose, ending in a broad flat swath across his mustache; his scraggly beard, probably grown to make him look older, since even in this photograph, taken when he was forty, Correns suffers the mixed blessing of being able to pass for a much younger man.

But on that weekend in April 1900, something in Correns had snapped. Meekness gave way to sarcasm, and he allowed himself some veiled accusations that stopped just short of charges of plagiarism. In his very choice of a title — "G. Mendel's Law Concerning the Behavior of Progeny of Varietal Hybrids" — Correns took away the priority claim of his own living rival, placing it squarely in the lap of a man who had been dead for sixteen years.

"The same thing happened to me which now seems to be happening to De Vries," Correns wrote, exposing his colleague in a manner so subtle as to be practically opaque. "I thought that I had found something new. But then I convinced myself that the Abbot Gregor Mendel in Brünn had, during the sixties, not only obtained the same result through extensive experiments with peas, which lasted for many years, as did De Vries and I, but had also given exactly the same explanation, as far as that was possible in 1866."

And isn't it a "remarkable coincidence," he dryly observed, that De Vries uses the same terms as Mendel — dominant and recessive — even though his citations do not mention the monk at all. He could have added an even more damning observation: that in all of De Vries's previous hybridization papers, and there had been several, never once had he used that terminology. Instead, he had always referred to traits as either "active" or "latent."

Correns took issue with De Vries's statement that "the hybrid shows always the character of one of the two parents, and that always in all its force; never is the character of one parent, which to the other is lacking, found reduced by half." To Correns the idea of "always" and "never" was ridiculous. He had already found many exceptions to Mendelian laws, and his experiments, though carried out with more plants than De Vries had used and for more generations, involved fewer species. If he could find exceptions on just a few species, how could De Vries have failed to do so in the dozen or so that he had crossbred?

Correns studied mostly *Zea mays* and *Pisum sativum,* and some of his findings were confusing. In crosses between maize varieties with either starchy or sweet kernels, for instance, he found that the Mendelian ratio for starchy/sweet hybrids applied in less than 5 percent of his crosses. And in his *Pisum* crosses, he observed something even stranger. When Correns crossed a yellow pea with a green pea, he did not get the expected F1 hybrid generation of all yellow peas. Instead, his F1 hybrids were almost transparent, resembling neither parent. No one knew at the time about the possibility, much less the existence, of a gene that codes for colorlessness, but that is the explanation of Correns's odd results. In this variety of *Pisum sativum*, both parents carried a gene that, when present (as it was in all the F1 hybrids), prevents the pea from showing any color at all.

Where Mendel had relied on the German word *Merkmal* to describe what was later translated as "factor" or "determinant," Correns used a better word: *Anlage*. The word makes clear, as neither *Merkmal* nor *Elemente* did, that Correns was thinking of the "determinant" as a discrete unit, a particle, something that could be transferred from parent to offspring. *Anlage* also carries associations of being responsible not for the characteristic itself, but for the code that leads to the characteristic. In retrospect, Correns's word *Anlage* seems close to the modern understanding of the gene.

Each trait has a single *Anlage*, Correns said, either dominant or recessive. Hybrids possess one of each form, in which case the dominant *Anlage* suppresses the recessive but does not destroy or change it. And, he said — in this instance going far beyond Mendel's original observations — an organism's entire set of *Anlagen* could be found in the cell nucleus.

Correns was the one who first used the 9:3:3:1 ratio to describe the results of dihybrid crosses. When he, and, earlier, Mendel, crossed *Pisum* that were double dominant for height and color (tall and yellow) with double recessives (short and green), then let the F1 hybrids self-fertilize, the F2 generation could be split into double dominants, double recessives, and — the great majority — mixed, that is, hybrid in one or both character traits. Correns categorized these types into four visually distinguishable groups, which fell out in the ratio 9:3:3:1. Although the ratio is often attributed to Mendel, the priest did not use it in his 1866 paper. It was Correns who came up with it and used it to support what he called "Mendel's laws."

Correns assigned to "Mendel's laws" names that the monk himself never used: the law of segregation and the law of independent assortment. But in his frenzy on that April night, Correns failed to differentiate between these two laws. It took an American geneticist, working more than a decade later, finally to separate the two concepts.

Correns recognized that the path to these laws was significantly easier in his day than it had been in Mendel's. Much had been learned in the intervening years, especially regarding the structure and function of the cell. This progress, Correns said modestly, meant that "the intellectual labor of finding out the laws anew for oneself was so lightened that it stands far behind the work of Mendel."

One of the big changes was in the general understanding of the cell nucleus. This had begun in 1869, shortly after Mendel was elected abbot. Johann Friedrich Miescher, a biochemistry graduate student from Basel, Switzerland, went about collecting bandages from hospitalized patients in postsurgical wards. As was common in that presterile era, the bandages were filled with pus. Miescher believed the white blood cells that proliferated in pus might hold some important secrets about the chemistry of the cell.

By haphazardly employing the most common laboratory techniques of the day — diluting his samples in sodium sulfide solution, in acid, in alkali, and in alcohol to see what precipitated out — Miescher hoped to find something of interest in the bandages. He collected an unfamiliar pure compound and decided that it had come from the nucleus of the cell. His next step was to try to isolate the nucleus, something that had never been done before.

Using dilute hydrochloric solution to destroy all the other cell parts, Miescher found, as he had hoped, that what remained were extracts of pure cell nuclei. He analyzed these extracts, looking for the same compound he had extracted from the bandages. He found that the compound made up a significant proportion of the nucleus. Ultimately Miescher found through biochemical analysis that the substance was composed of nitrogen, phosphorus, and chromatin in an acidic solution. He called it nuclein, after the nucleus.

But, like Mendel before him, Miescher never learned the significance of what he found. He died from a chronic chest ailment in 1895 at the age of fifty-one, before anyone realized that nuclein was really DNA, the most important chemical on earth, the compound that transmits, from parent to child to grandchild, every bit of information needed about every living thing.

The third supposed rediscoverer to emerge that spring was Erich von Tschermak, a twenty-six-year-old graduate student who was crossbreeding peas and also wallflowers in Ghent, Belgium, and in Vienna. His part in the rediscovery is usually dismissed these days in a paragraph or two. Though he worked hard to be called a rediscoverer, few people even then believed that he really understood Mendelian ratios, the notion of dominance, or the implications of segregation.

Tschermak was a grandson of Eduard Fenzl, the Viennese botanist with whom Mendel is said to have feuded in 1856 during his doomed second try for a teaching certificate. Despite his youth, Tschermak had his eye on posterity. "It was not easy for [me] to establish [my] part in the discovery of Mendelism and in its utilization for practical breeding," he wrote years later, "for only the names of De Vries and Correns found a place in the then leading textbooks. This omission, however, was corrected in the later editions." But it was quickly reconsidered, as historians of science began to question Tschermak's position as a rediscoverer. His paper, published a month or two after the others in June 1900, presented his findings tentatively; he made no attempt to deduce general principles from his experimental data. He failed to see that hybrids carry pairs of determinants that differ from each other, that the pairs segregate in the gametes, and that this behavior accounts for predictable ratios in different crosses. The consensus today is that Tschermak was a dutiful graduate student who managed to find and cite Mendel's paper but who never quite understood the essential points of Mendelism.

When William Bateson returned home to Cambridge in July 1899 after the International Conference on Hybridization, he was so impressed with Hugo De Vries that he took it upon himself to spread his gospel throughout England. He arranged to speak again before the Royal Horticultural Society the following spring, hoping to captivate its members — and to spur them on to crosses of their own — with a description of De Vries's mutation theory.

The RHS was a surprising venue for such a challenge. The society was mostly a gentleman's association without much scientific muscle. Its main activities were social, such as the annual Temple Gardens Flower Show, held each May and described as "one of the principal events of the London season." The flower show drew such dignitaries as the queen of Sweden and Norway, the duchess of Connaught, the duchess of Devonshire, Lady Warwick, and Lord Cross. But the RHS was sponsoring a new scientific section, under the direction of Maxwell Masters, in an attempt to gain some intellectual credibility. The transition was a halting one: the night after Bateson's scientific lecture, members of the society gathered on more familiar terrain, at the lavish annual dinner of the Royal Gardeners' Orphan Fund.

No one knows for sure what Bateson did on his way to London on the morning of Tuesday, May 8, 1900. The most familiar story is a vivid and appealing one: that in preparation for his RHS lecture about the mutation theory, Bateson had read the version of De

Vries's paper, published by the German Botanical Society in its *Berichte* of April 25, and had tracked down Mendel's paper, cited by De Vries in that famous footnote. During the one-hour train ride into London on the very morning of his RHS address, the story goes, Bateson read Mendel's paper and was so struck by its brilliance and clarity that he completely revised his speech, and used the occasion of his lecture to introduce Gregor Mendel to the English-speaking world.

But such a scenario requires several things to have happened — each of them possible but all of them unlikely. Assuming that he managed to get his hands on De Vries's article and read it carefully no later than the weekend before his lecture, Bateson would have had to go to the university library sometime on Monday, May 7, to find the journal mentioned in De Vries's report, the one with Mendel's paper. The library did indeed have a bound volume of the Brünn *Verhandlungen* from 1866. But could Bateson have obtained it in time to catch his Tuesday morning train?

Some people think he could not have, or at least that he *did* not. For one thing, the report of Bateson's lecture written by Maxwell Masters for the May 12 edition of the RHS's weekly *Gardeners' Chronicle* makes no mention whatever of Gregor Mendel. This suggests that Bateson spoke on May 8 only about De Vries, not about Mendel. He may have been reading de Vries's *Berichte* paper on that train ride or, more likely, his earlier, shorter paper from *Comptes Rendus* — which did not cite the Moravian monk. It may have been days or weeks after Bateson's return to Cambridge — at the height of springtime, when all his plant and animal experiments required extra attention — before he finally came across Mendel's paper for the first time.

The most interesting question in this game of academic shuttle-cock is why it all mattered so much to Bateson. Why did he need to perpetuate the story of his epiphany on the train to London? Why did he always say, as his widow repeated in her memoirs, that his recognition of Mendel's brilliance came to him in a flash? It seems to have been part of the mythology — the mythology of the genius of both Mendel and Bateson himself.

Bateson's inability to hold two competing thoughts at once, his tendency to see the world in stark blacks and whites, drove much of the debate in the years after Mendel's rediscovery. To Bateson a person was either grand beyond measure or insignificant, either his best friend or his most bitter enemy. In order for Mendel to deserve the title of father of a new science, in this either-or mentality, he needed to have been the kind of genius who had eureka moments that ordinary mortals never do. However, Mendel's particular brilliance was something quite different, not at all the glittering type that suited Bateson's story. But the story persisted and grew. How nicely it worked, in terms of the narrative, for the flamboyant genius of the Mendel of legend to be echoed, thirty-five years later, by another eureka moment on a southbound Great Eastern Railway train heading into Liverpool Street Station, London.

Whether Bateson read the 1866 paper on that train or weeks or even months later, from that moment on he became Gregor Mendel's chief apostle. He had the paper translated into English and published it in 1902 along with a spirited, and somewhat acerbic, preface called *Mendel's Principles of Heredity: A Defence.* In it he criticized the biometricians, who believed, following Darwin, in slow, continuous evolutionary change. Bateson and his supporters, whom he had taken to calling the Mendelians, believed that evolution occurred as a result of large "discontinuous" changes from one generation to the next; they said Mendel's work with peas revealed how large-scale evolutionary leaps might occur.

In addition to this split over the pacing of evolution, the biometricians and the Mendelians differed in other ways. The biometricians, for instance, placed great stock in statistics to analyze Darwin's theory of natural selection. Mendelians — and this was ironic, given Mendel's own predilection for mathematical analysis — dismissed statistics, preferring to find evidence of evolution through empirical research. The two groups had the same ultimate goal of making biology an exact science, but they disagreed about how to achieve it. "Exactness is not always attainable by numerical precision," Bateson wrote in a pointed reference to the biometricians: "there have been students of Nature, untrained in statistical nicety, whose instinct for truth yet saved them from perverse inference, from slovenly argument, and from misuse of authorities, reiterated and grotesque."

De Vries, for his part, grew less and less impressed with Mendel as the years went on. This change of attitude may have arisen from pure and simple envy — envy of a man who had been dead for a generation, yet who had managed to steal De Vries's very reputation. He had meant to use Mendel's findings only as further support for his own mutation theory, a steppingstone to what he considered a far more meaningful approach to the study of inheritance and evolution. How frustrating it was to watch the steppingstone quickly be turned into a new — and ultimately competing — theory. De Vries was "quite jealous of the rapid development of Mendelism," observed Tschermak years later, "and considered his mutation theory somewhat ignored, especially among breeders." Tschermak thought jealousy the only possible motive for De Vries's failure to mention Mendel in his 1907 book *Pflanzenzuchtung* ("Plant Breeding"), and for his "brusque refusal" to sign Tschermak's petition in 1908 supporting a commission to create a Mendel memorial in Brünn.

De Vries believed Bateson's enthusiasm for Mendel was misplaced. "Please don't stop at Mendel," he wrote a year and a half after the 1900 spring of rediscovery. "I am now writing the second part of my book [*Die Mutationstheorie*] which treats of crossing, and it becomes more and more clear to me that Mendelism is an exception to the general rule of crossing. It is in no way *the* rule!"

But Bateson paid him no mind. He simply crossed De Vries off the ever-shrinking list of the people who were one of "us." If not for Bateson, Mendel's pea experiments might never have become the unifying starting point of genetics. By the same token, if not for Mendel, we might know — or care — very little indeed about the opinionated zoologist from Cambridge.

16 ✺

The Monk's Bulldog

*Gardeners are the ones who, ruin after ruin, get on
with the high defiance of nature herself, creating,
in the very face of her chaos and tornado, the bower
of roses and the pride of irises.*

— *The Essential Earthman*,
Henry Mitchell, 1923–1993

THE LECTURE WAS NOT SCHEDULED to begin for another hour, yet the room was packed. Every seat in the hall had long been filled; now people were beginning to perch on windowsills or lean into out-of-the-way niches along the walls. Rumors of an impending fight had been rumbling all morning, bringing out both the intellectually engaged and the morbidly curious. No doubt the fight would appear tweedy and civilized, in the fussy, carefully clipped manner of British academics. But who wouldn't want to see a good healthy row, even one you had to infer between the lines?

William Bateson was first to stride onto the stage, his long legs catapulting his six-foot-plus frame as though they were attached to someone else's body. If Friday, August 19, 1904, was a day like any other, Bateson was wearing a wool gabardine suit, a high-buttoned vest, and a white shirt with a tall, tight collar. If it was a day like any other, he had a copy of Voltaire's *Candide* tucked into a pocket — his wife and frequent collaborator, Beatrice Bateson, used to say that her husband almost never went anywhere without it. If it was a day like any other, his huge mustache was drooping over his full lips, his eyebrows were bristling, and he was primed for a fight.

And if it was a day like any other, this imposing, energetic bear of a man was suffering from stage fright. This always afflicted him, whether he was about to speak to a class of Cambridge under-graduates, a group of ladies from the local horticultural club, or, as on this day, a group of partisan listeners crammed into a hot lec-ture hall to await the scholarly equivalent of a Sunday afternoon bullfight.

Nervous as he was, Bateson was on home ground. This meeting of Section D, the section on zoology, of the British Association for the Advancement of Science was being held in Cambridge, the university town in which Bateson had been born, raised, married, and where, at the age of forty-five, he still taught and conducted research. More to the point, he had just been installed as the new president of Section D.

These were his people. Unless, that is, he had completely alien-ated them the morning before. On Thursday, August 18, Bateson had delivered a presidential address to the section that probably caused more uncomfortable shufflings than had any other address in the association's seventy-three-year history. In a diatribe un-characteristic for staid British academia — but perfectly in keeping with Bateson's own pugnaciousness, and with the rancor of the debate in which he was embroiled — Bateson had used his presi-dential prerogative to demean the ideas and methods of his arch-rivals, the biometricians. He had done this time and again for four years, ever since assuming the role of chief apostle for Mendel's theories and taking on the mission of bringing the monk's work to the attention of the English-speaking world. He derided the bio-metricians' most impressive statistical feat, the correlation table, as a rigid structure in which "the biometrical Procrustes fits his ar-rays of unanalyzed data." The table, a complex gradient of inher-ited traits and their correlation with those traits in earlier genera-

Nature of group dealt with	Correlation of parent plant and offspring capsule	Regression of offspring on parent plant
(a) Early capsules (apical flowers) of principal plants	.2323	.4003
(b) Capsules on plants — not starvelings, i.e. with at least three capsules	.2430	.4050
(c) All capsules on principal plants	.2295	.4295
(d) All capsules on all plants	.1960	.4064

A correlation table of the biometricians.

tions, might have looked "imposing," Bateson intoned, but it was "no substitute for the common sieve of a trained judgment."

The fight was the one between "continuity" — the biometricians' view of evolutionary change — and "discontinuity" — Bateson's view. Bateson had spent the previous four years rallying his camp around the nineteenth-century genius whose paper he had discovered on that train ride to London (or shortly thereafter): Gregor Mendel.

In his presidential address, Bateson found fault not just with the biometricians but with the entire scientific branch of zoology. Zoologists had become mere cataloguers, he said, having wandered far from their proper task, which was to find "the fundamental nature of living things." But with the job of describing all the animals on earth "happily approaching completion," Bateson said it was time for zoologists to embrace the future by shifting their attention back to the central questions of the twentieth century, those of heredity and evolution.

Now, on Friday afternoon, the second half of his planned one-two punch against the biometricians was about to begin. Bateson had choreographed this preemptive strike since early June. Every morning throughout the summer, he had taken a lawn chair and table out to the grove of shrubbery behind the chicken coops at his home in Grantchester, a village just beyond the Cambridge city limits. Each day's work began with his assistant Reginald Crundall

Punnett, a fellow zoologist at Cambridge University, bringing the *Morning Post,* the only local newspaper Bateson read, and then only for news about art auctions and sales. After a quick glance at the paper, Bateson set to work writing, rewriting, rethinking, and revising his presidential address and the following day's presentations. As the older man worked, leaving strict orders not to be disturbed, Punnett set about attending to the sweet peas with which he was replicating and enlarging upon Mendel's experiments. While Punnett recorded and labeled peas, Bateson would occasionally emerge from his little work station in the glen to casually ask Punnett's opinion of one point or another. But he already knew what he wanted to say; indeed, he already knew how he wanted to say it. He brought things up with Punnett in part just to hear himself speak.

Throughout the summer Bateson relished the upcoming fight with the biometricians much as he relished a good tussle on the croquet field. (Inexplicably, Bateson always wore a fez when he played croquet.) Years later the novelist Nicholas Mosley described a character closely modeled on Bateson, whose most salient feature — besides his brilliance and his determination to be Mendel's mouthpiece — was his single-minded aggressiveness on the playing field. For the fictional Bateson, the game of choice was tennis rather than croquet. He "used a tennis court as some sort of battleground on which to engage with the people (and these seemed to be most people) against whom he felt aggression. He . . . put great energy into his game; he would serve and rush to the net; he would leap to and fro volleying; he would prance backwards towards the baseline slashing at high balls as if they were seagulls."

In real life Bateson could indeed be a ferocious opponent — as he demonstrated on that hot Friday in August 1904. He stacked the morning's program with presentations by his allies. Miss Edith Saunders, his long-time research assistant, spoke about their experiments on *Datura stramonium* (jimsonweed), *Silene alba* (catchfly), *Mathiola incana* (gillyflower), and other plant species. Colonel C. C. Hurst talked about his work with rabbits, such as his crosses between Belgian hares and inbred Angoras. And A. D. Darbishire, who until just months before had been on the side of the biometricians, focused on animal experimentation, specifically on his work crossing albino mice with Japanese waltzing mice.

Having Darbishire on his side was a major coup for Bateson, which he had accomplished by scaring the wits out of the younger man. Darbishire had been a promising protégé of Walter Frank Raphael (known to his friends as Raphael) Weldon, the leading advocate of the biometricians. In the previous year and a half, Darbishire had published four anti-Mendelian papers in *Biometrika,* a journal cofounded by Weldon. Darbishire's crosses between albino mice — with the recessive traits of white fur and pink eyes — and waltzing mice — with a recessive trait that made them spin in circles when other mice stood still — showed no support for the Mendelian ratios, the young man concluded. Instead the hybrids confirmed Galton's theory of ancestral heredity, which said that no inherited contribution from an ancestor, however distant, is ever lost.

Bateson, suspicious of Darbishire's results, wrote to him seeking his original data. In May 1904 he realized that Darbishire's data were inaccurate, incomplete — and possibly falsified. He revealed this in a letter to Darbishire on May 22, to which Darbishire replied immediately with a desperate plea for discretion — something Bateson had in only limited supply.

"I hope you will do your best to get me out of the position I am in as soon as possible and I pray you not to mention this letter to anyone," Darbishire wrote. "What do you suggest?"

What Bateson suggested was total — and public — capitulation. Less than three months later, when Darbishire spoke at the British Association meeting, he rejected all prior interpretations of his findings and said that they proved, rather than disproved, Mendel's laws. The biometricians denounced Darbishire, who devoted some effort to attempting a reconciliation between the two camps — and then, failing that, became a devoted and lifelong Mendelian.

At last, at about the time the sun was surely baking the roof of the new Sedgwick Museum of Geology, and the people crammed into the museum's lecture hall were doubtless fairly roasting in the stifling air, Bateson brought to the podium Raphael Weldon, his most vociferous opponent. In their youth, Weldon had been Bateson's best friend. Now, in middle age, he was his bitterest and most impassioned enemy.

Weldon was a volcanic speaker. He gestured so violently when he was excited, as he was that Friday morning, that the sweat glistening on his bald dome danced off his head with every flourish and dripped from his brow and cheekbones onto his prepared comments. Weldon used his standard arguments against Bateson and the Mendelians: the occasional appearance of remote ancestors' traits, known as atavism; the intermediate nature of some hybrid crosses; the possibility that the Mendelians' results could be explained with other hypotheses. He called their methods careless and their theories about underlying mechanisms "cumbrous and undemonstrable."

A slight, pale man with a severe mustache and a fringe of dark hair, Weldon often lost his composure when he felt strongly about a cause — to the detriment of his own arguments. But his Cambridge audience on this August morning was moved by both his volubility and his intellect. "Clever beggar, that," one young Oxford student said to his friend with admiration; "he hasn't got to stop and think."

After Weldon's talk, Section D broke for lunch. This was no doubt part of Bateson's strategy. He was, after all, a chess player, having taken up the game when he learned that Mendel had been an avid player himself. (For the same reason he had taught himself to smoke cigars, and subscribed for a while to *Die Fliegende Blätter,* Mendel's favorite humor publication.) Bateson probably thought it best to allow time for his audience's brains and body temperatures to cool down before taking his own turn. He might also have hoped that interest in the afternoon session would grow during the lunch break, as men murmuring in corridors and nearby cafés informed other British Association members about the great fight brewing in the zoology hall.

The afternoon session started off much as the morning session had — with reports of experiments from some of Bateson's minions. Punnett described his and Bateson's work on fowl, and Minot talked about his experiments on guinea pigs. Then it was time to hear from the president. The event had been so carefully orchestrated that the lecture hall into which William Bateson strode was exactly as he had pictured it: crowded, hot, and as expectant as a child on Christmas morning.

Bateson was an arrogant man who fiercely defended his supporters and just as fiercely trounced his enemies. And Weldon had tumbled far in Bateson's hierarchy of allegiance.

In the early 1880s, while they were both at St. John's College at Cambridge, the two had been the best of friends. Bateson at the time cut an awkward and gangly figure; his very movements were "unconventional," a classmate later recalled, his whole aspect "a living protest against the 'average.'" He was large, untidy, and out of control. Weldon, in contrast, always looked well-trimmed and carefully turned out, his slim bearing and pale, fine skin a testament to the frail health that would ultimately lead to his unexpected death. How strange they must have looked together, traversing the crisscrossing paths of the courtyards of St. John's, their student robes flapping in a breeze freshened by the River Cam.

Although Bateson dominated in physical bearing, Weldon was the spiritual leader of the twosome. It was he who suggested that his friend embark on what would be a turning point in Bateson's scientific career: a trip to America during his first two summers after graduation. He thought Bateson would benefit from working with William K. Brooks, a morphologist at Johns Hopkins University interested in the path of evolution from invertebrates to the more complex vertebrates. Brooks was studying an animal that would become Bateson's first research model, a worm known as *Balanoglossus,* which lived in the warm waters of Maryland's Chesapeake Bay.

Weldon was only a year Bateson's senior, but he was, as an undergraduate and for the rest of his life, more entrenched in the old-guard academic elite. He achieved a professorship at University College, London, by the age of thirty, which thoroughly rankled Bateson, who was not made a professor (at Cambridge) until he was forty-seven. Weldon was such an overshadowing presence that Bateson had been heard to complain that when the two first knew each other, even though they were friends, he was often made to feel "like Weldon's bottle-washer" — the job in any laboratory given to the lowliest person. Maybe Bateson was always aware of being, as the British put it, "one generation from trade." His father rose to become the master of St. John's College, making him a highly respected man in Cambridge, but both grandfathers had been Liverpool merchants. Weldon, too, was the grandson of a middle-class manufacturer, but his father, who started out as a journalist, became rich by inventing a new chemical process for making chlorine. His financial comfort might have given Weldon the air of an aristocrat.

The first years of Bateson's research adventures overlapped the final years of Gregor Mendel. In the summer of 1883, when Bateson went to Maryland to work with Brooks, Mendel was still fighting the monastery tax, still taking his weather readings, still playing Sunday afternoon chess with his nephews. By Bateson's second American summer, in 1884, Gregor Mendel had been dead for nearly six months.

At first Bateson, like Brooks, was concerned with morphology — the study of the forms of animals. He performed his work with *Balanoglossus* so well that he discovered new evidence that the worm was an intermediate between invertebrates and vertebrates. In the next two years he published three major papers on the evolutionary significance of *Balanoglossus* in the *Quarterly Journal of Microbial Science* — a remarkable feat for a man so young. Then he set off on an extended and difficult trip to Russia, hoping to find more evidence of large-scale variations among different plant species growing under different conditions in the salt lakes scattered across the steppes. He spent eighteen months away, miserable and alone, wintering in the village of Kazalinsk in Turkestan and summering out near the lakes. During that time Bateson decided morphology was an outdated field.

He saw an analogy with the steam engine, just then transforming the transportation industry. "Presently steam will be introduced into Biology," he wrote to his mother in 1886, "and wooden ships of this class won't sell." The biological steam engine, of course, was understanding how organisms change, not merely describing how they were at present, which the "wooden ship" of morphology was chiefly concerned with. In short, the future of biology would be filling in the details of evolution.

While Bateson was going through this crisis of the mind, his friend Weldon was back in England going through some crises of the heart. Weldon's emotional blows began several years earlier, while he was still an undergraduate at Cambridge. In June 1881 his younger brother, Walter Alfred Dante Weldon, who had joined Raphael at Cambridge, had dropped dead of apoplexy at the end of his first semester. He was nineteen years old. Within weeks Weldon's grieving mother was dead, and the following summer

Raphael's beloved mentor, the esteemed biologist Francis Maitland Balfour, died while mountain climbing. Loss, then another loss, and another. And in 1885, four years after the death of his brother, and two years after his marriage to Florence Tebb, Weldon lost his father, who died unexpectedly at the age of fifty-three.

Weldon kept these tragedies to himself but was forever haunted by them. As his close friend Karl Pearson, a brilliant mathematician and statistician at University College, London, put it, Weldon's outgoing manner belied an underlying "tinge of melancholy, a doubt whether he too would live to finish his work, and a tendency to take the joy and fullness of life while it was there." Weldon's lows were very low, but his highs were very high. Like Bateson, he brought an "almost boyish delight" to his work and seemed blessed by boundless energy. But he was more approachable than his stormier, more opinionated friend. Weldon debated with colleagues far into the night, gave vigorous classes and public lectures, and bicycled as much as one hundred miles in a day. "To see Weldon keen over a piece of work was to believe him robust and ready for any fray," said Pearson; "but looking back on the past one can see what each piece of work cost him."

After Bateson came home from Russia, he spent the next seven years working on his first book. "My brain boils with Evolution," he wrote to his sister Anna. During this time he and Weldon were drifting apart, finding themselves on opposite sides of the debate over evolution's pacing. Bateson was becoming a diehard discontinuitarian; Weldon, with his new friend Pearson and their mentor, Francis Galton, came down on the side of continuity. On this question Galton was something of an anomaly; his statistical proofs were used as support for both the continuous and discontinuous camps, and each side claimed him as a spiritual leader — competing claims that Galton did nothing to oppose.

The split between the two groups turned ugly in the spring of

1894, when Bateson's long-incubating book was published and Weldon publicly embarrassed him by writing a critical review. The book, *Materials for the Study of Variation: Treated with Especial Regard to Discontinuity in the Origin of Species,* described in vivid detail nearly nine hundred examples of discontinuous variation. Weldon's review, for the journal *Nature,* was one of history's most dramatic confirmations of the old publishing adage: friends should not review each other's books.

Weldon began by praising the "descriptive" first half of *Materials,* which he said "must be carefully read by every serious student; and there can be no question of its great and permanent value, as a contribution to our knowledge of a particular class of variations, and as a stimulus to further work in a department of knowledge which is too much neglected." But Bateson — like most authors — focused only on Weldon's criticism. The second half of the book contained "several inaccuracies," Weldon wrote, "due partly to want of acquaintance with the history of the subject." Bateson's central point was that discontinuous variation was essential for creating new species. Weldon's central point was that it was not.

When the critique appeared in *Nature* on May 10, Bateson reacted venomously. As he confessed later in one of his lengthy missives, in handwriting so insistent that at times it could scarcely be read: "If ever a man set himself to destroy another man's work, that did he do to me."

Maybe Bateson took such personal affront in part because his heart was already sore from lovesickness. The woman he adored, Caroline Beatrice Durham, had been kept from marrying him four years earlier — not because she didn't love him (she did, madly) but because her mother felt Bateson had drunk too much wine at the couple's engagement party. Mrs. Durham was especially sensitive to this because Caroline's father was a secret alcoholic. Bateson

had nursed his broken spirit by immersing himself in his work —
and now the work, too, was being rejected, by a man he felt closer
to than almost anyone else he knew.

Nine months later Bateson's hurt feelings led him into a spirited
public controversy, not with Weldon but with a surrogate. At a
meeting of the Royal Society on February 28, 1895, a biologist
named W. T. Thiselton-Dyer gave a talk about the origin of new
hybrids of cineraria, a hairy-leafed perennial, properly called
Senecio cruentus, the flower of which has a profusion of tiny bris-
tles that give the genus its name, the Latin for "old man."
Thiselton-Dyer reported that the wild type of *S. cruentus,* found
on the Canary Islands off the coast of Spain, differed dramatically,
in flower shape and color, from a recently cultivated form at the
Royal Gardens at Kew. These differences within the same species,
he said, were proof that natural selection worked through small,
continuous changes. To Thiselton-Dyer, the hand of the breeder at
Kew, creating artificial selection, was analogous to nature's hand in
natural selection. In both instances, small differences were sufficient to lead to big changes — resulting in new varieties and,
given enough time, entirely new species.

Thiselton-Dyer published these thoughts in a letter to the editor
of *Nature.* Bateson wrote a letter in response, thrashing his opponent for his "misleading" statement, which he said "neglects two
chief factors in the evolution of the Cineraria, namely, hybridization and subsequent 'sporting.'" Sporting was the term first used
by Darwin's critic Fleeming Jenkin to mean the occasional eruption of sudden changes in a species' makeup that lead to dramatic,
unexpected differences.

Over the next two months, ten letters about cineraria appeared
in *Nature* — from Bateson, from Thiselton-Dyer, from Bateson
again, and at last from Weldon. Bateson took special offense at
Weldon's letter of May 13, 1895. "All I wish to show is that the

documents relied upon by Mr. Bateson do not demonstrate the correctness of his views," Weldon wrote, "and that his emphatic statements are simply of want of care in consulting and quoting the authorities referred to." The accusation of carelessness, made by one conscientious scientist about another, was particularly wounding.

Hoping for a rapprochement, Bateson and Weldon met in person on Tuesday, May 21. The meeting did not go well. Weldon excused Thiselton-Dyer's arguments as just so much "bluffing." If this was the case, Bateson parried, then Weldon must have been "the accomplice who creates a diversion to help a charlatan." Faced with such accusations, Weldon gave up. On Friday he fired off a disgusted letter that began "Dear Bateson, I can do no more." The chance for civility had long passed, he told his former friend: "If you insist upon regarding any opposition to your opinions concerning such matters as a personal attack upon yourself, I may regret your attitude, but I can do nothing to change it."

The letters to *Nature* continued. Thiselton-Dyer accused Bateson of "facile theorizing" and "barren dialectic." Bateson said Thiselton-Dyer was theorizing, too, and without any facts to back him up. Finally, the journal editors had had enough. In June 1895 they refused to publish another sentence on the subject.

But the damage had been done. Bateson and Weldon never again spoke a civil word to each other.

While his affection for his erstwhile friend and colleague was disintegrating, Bateson's other affair of the heart was taking a dramatic turn for the better. In September 1895 — soon after the cineraria controversy had played itself out in *Nature* — another controversy was being resolved in the pages of a very different journal: the *English Illustrated Magazine*, a popular woman's monthly. In a short story under the byline of Beatrice Durham —

the young woman to whom William Bateson had been engaged six years before — a mousy middle-aged woman named Sophy accompanies her beautiful young niece to a ball and stuns the niece by capturing the eye of the evening's guest of honor, Sir William Collins. On the carriage ride home, the niece discovers that Aunt Sophy and Sir William loved each other in their youth, and that their plans to marry were thwarted by Sophy's mother. In the intervening years Will has turned his life around. He has gone on "expeditions," been knighted for his service to the Crown, and finally has shown up again before her, resplendent in top hat and tails. Aunt Sophy, now in her forties, having never recovered from the heartache of losing her first beau, falls in love all over again. He does, too.

"Will you marry me?" Sir William asks Sophy. "No, no, William," she replies heatedly, "it's too late, dear — too late now." Well, then, Sir William asks "huskily," may I call on you, so that we can try to get to know each other once again? "Oh, William," she exclaims, "if you will!"

This story was meant to be a message to Bateson from his Beatrice, the middle name she used then and forever after. I regret having lost you, the message was meant to say; all obstacles have been removed now that my parents have died; I long to see you again. In 1895 a maiden would never be so direct as to send a letter bearing such a message or to utter these words aloud. Indeed, it was forward of Beatrice to announce herself even in this oblique fictional way. But the effort would have been in vain had not the wife of the famous Cambridge philosophy professor Alfred North Whitehead, who was a close friend of Bateson's, taken the initiative to send him the September issue of the *English Illustrated Magazine* the following April. Bateson, of course, had never seen it — he was not in the habit of reading women's magazines — and he instantly dispatched a letter to Miss Durham.

"I have been led to think it possible that you may be willing to see me again," he wrote in his earnest scrawl. "If it is not so, you tell me and that will be all; but if it is so, will you some day meet me? . . . [I]t has been for a long time my earnest desire to meet you again, if only as one who was once my dear friend, without regard to the future at all." But the future quickly announced itself anyway. In a matter of weeks the two were engaged, and on June 16, 1896, they married at last.

Bateson began to build the family he had always wanted. In 1898 Beatrice gave birth to a son, John, and the following year another son, Martin. The four of them were bursting out of Norwich House, their shallow three-story stone house on the corner of Norwich and Patton streets, with tall bay windows and a wrought iron gate enclosing its tiny thumbprint of greenery.

In 1899 Bateson moved his young family to Grantchester. The walk to the tiny village from Cambridge winds along the stone-lined River Granta, an easy two miles punctuated by footbridges and meadows. The village itself is not nearly as charming as the walk — just a few streets and high hedged walls that hide the more impressive houses from the curious eye. Behind one of these hedges, on a winding lane called Broadway, just down the road from the Rose and Crown pub, is Merton House, where the Batesons lived for eleven years. The property included orchards, a garden, several outbuildings, and an enclosed paddock.

"Would I were/ In Grantchester, in Grantchester!" wrote the village's most famous native son, the poet Rupert Brooke. Sitting in the Café des Westens in Berlin on the eve of World War I, Brooke was filled with nostalgia for a place where he could,

> *flower-lulled in sleepy grass,*
> *Hear the cool lapse of hours pass,*
> *Until the centuries blend and blur*
> *In Grantchester, in Grantchester . . .*
> *Say, do the elm-clumps greatly stand*
> *Still guardian of that holy land? . . .*

Stands the Church clock at ten to three?
And is there honey still for tea?

Brooke would never see Grantchester again. He died of blood poisoning while his navy ship was stationed in the Mediterranean. In his honor the Rose and Crown was renamed the Rupert Brooke Pub. And when the church clock was repaired in 1985, the hands were left frozen for all time just the way Brooke remembered them — poised at ten minutes to three.

Merton House offered rather primitive conditions for Bateson's numerous experiments. The problem was the scarcity of money rather than the cramped quarters — though cramped they still were, especially after the birth of Gregory, named after Gregor Mendel, in 1904. Through an artful exchange of what he called "begging letters," Bateson managed in the last days of 1903 to secure £150 a year for two years from his friend Christiana J. Herringham. Mrs. Herringham acknowledged that the money "isn't as much as you want," but said it might pay for a research collaborator and might help Bateson not "hanker after America." He was, indeed, flirting with the idea of moving to America, where he had been gloriously received on a lecture tour the year before, and from which he had already received one job offer, from Dr. A. G. Mayer of the Brooklyn Institute. After receiving Mrs. Herringham's support, Bateson promptly wrote to Leonard Doncaster, a biologist from Cambridge, asking him to join him as a full scientific collaborator. Doncaster declined, saying he preferred to work on his own — though for many years he counted himself among Bateson's acolytes, conducting experiments that Bateson would use to support his case for discontinuous variation. Bateson then tried his second choice. On Christmas Day of 1903 he wrote to R. C. Punnett, a zoology demonstrator at Cambridge, asking him to enter "into partnership in my breeding experiments. . . . We could do far more in combination than separately."

Punnett, a well-liked young man, was at the time more famous for his athletic brilliance (in cricket, golf, tennis, and a British game called fives) than for his scientific acumen. He accepted the job with enthusiasm but declined the £80-a-year salary Bateson offered. He had an independent income from the wealthy Punnett family of Lincolnshire fruit growers — a business so famous that its very name entered the British vocabulary, "punnett" being the word for a large wicker basket used for gathering strawberries.

The Punnett name is also familiar today, especially in the United States, for quite another reason, one directly related to R.C. himself. It was he who invented the easiest way to visualize the crosses Mendel described in his paper of 1866. Mendel had used a few charts showing the possible combinations for each of the four gametes involved in a monohybrid cross, and his successors had tried to refine those charts to make the combinations more clear. But Punnett's matrix was the clearest. He called it a checkerboard, setting it up so that the female contribution was shown in the horizontal boxes and the male contribution in the vertical boxes. In each box, then, two gametes intersected — one from the top, one from the side — and it was an easy matter to write in the box which two gametes had paired to form each particular offspring.

Punnett's matrix first appeared in the third edition of his textbook, *Mendelism,* in 1911. Not until his death in 1967, at the age of ninety-one, was the checkerboard renamed the Punnett square. As a result, Punnett's name is far more familiar to casual genetics students, and to the general public, than is Bateson's.

Bateson plunged into his collaboration with Punnett as ferociously as he approached every other venture, from playing croquet to collecting Japanese lithographs to reading passages of the Bible to his sons every morning. He seemed to be everywhere at once, overseeing projects in every corner of his property — and beyond. "The poultry occupied a small paddock split up into about two dozen

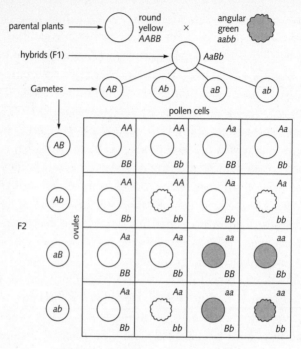

A Punnett square showing a dihybrid cross.

pens," recalled Punnett about the working conditions of 1904, during the time Bateson was composing his presidential address for the British Association. "There were several incubators in a bedroom upstairs though this had soon to be given up since it was requisitioned for the little boys' governess. The chicks were reared in movable brooders along the garden paths. It was not a very satisfactory arrangement for, in a wind, one of them occasionally caught fire [the brooders being warmed by oil lamps], and there was an end to *that* hatch."

Despite his curmudgeonliness, Bateson attracted a constant flow of research assistants, many of whom remained devoted to him for years — even though they were forced to work in whatever nooks and niches they could find. Many were young women who taught or studied at one of the two women's colleges within Cambridge University — Newnham or Girton. With so many women assistants, it seems inevitable that some were in love with him. That certainly may have been the case with his first assistant, Edith Saunders, a lecturer in botany at Newnham who joined Bateson's circle in 1894 and never left. Why else would she have stuck for so long with a man who treated her coldly, took her scientific partnership for granted, and occasionally made fun of her stiff, mannish bearing and her formidable manner? When "Mr. Bateson" and "Miss Saunders," as they always called each other, together received Hugo De Vries at the Bateson house in 1899, Bateson wrote an amused letter to Beatrice, who was vacationing with the children, about Miss Saunders's uncharacteristically flirtatious behavior with the courtly Dutchman. She "talked and chattered as I never saw her do before," he wrote. Perhaps he would have seen her flirt and chatter more often if he had ever taken note of her.

Miss Saunders bred *Biscutella laevigata* (the Cruciferous Spectacle plant) on land behind the Cambridge University Botanic Garden that the Batesons had rented when they lived in town. Even after the family moved to Grantchester, she continued her experiments in that familiar "allotment garden." In a field behind Newnham College, two other women who worked for Bateson — Miss Sollas and Miss Killby — raised guinea pigs and goats. Florence Durham, Beatrice's older sister, was another assistant, consigned to a museum attic to crossbreed mice. Back in Grantchester, Miss Muriel Wheldale grew snapdragons *(Antirrhinum)* and Miss Dorothea Marryat, four o'clocks *(Mirabilis jalapa)*. Others, including some men, worked here and there around Oxfordshire; Mr. Staples-Browne bred pigeons, Major C. C. Hurst bred poultry and rabbits, Leonard Doncaster bred Abraxas moths, and Miss Nora Darwin grew wood sorrel *(Oxalis)*.

Bateson's most loyal assistant was his wife. While raising their three sons, Beatrice also spent long hours helping with the hard chores involved in maintaining a rotating stock of experimental plants and animals. She was a full partner in both manual and intellectual labor, and she recorded the difficulty of their tandem ventures without any apparent trace of bitterness or regret. "From the merest menial drudgery to high flights of scientific speculation, hand and brain were hard at work," she wrote in a memoir published in 1928, two years after Bateson's death. "There was all the sorting, sowing and gathering of seed to be done personally; the fertilizing and recording; most of the digging, hoeing, weeding, staking and watering; the five incubators, each 100 egg power, and as many rearers (all run with oil lamps); the tiny chicks; and at times hundreds of larvae to be attended to. All writing (not reckoning the ordinary post, which was often heavy) was done at night."

Because of the constant pull of chores, the Batesons took separate vacations, so that one of them was always home to collect the eggs and feed the animals. Between early spring and late autumn, the only time they left the house together was for their annual foray into London to see the flower show of the Royal Horticultural Society in Temple Gardens — the exposition that *The Times* of London called the social event of the season. Then they made a day of it, visiting art galleries or auction houses — Bateson was a fervent collector of old masters' drawings and Japanese prints — or the Tate or some other museum. In his later years, Bateson sat on the board of directors of the British Museum, which gave him more pleasure than all the accolades he had by then accumulated, including the Darwin Medal, the Royal Society Medal, and honorary membership in the Brünn Society for the Study of Natural Sciences.

Early in their marriage the Batesons occasionally managed to squeeze in an evening of theater in the West End. But once they moved to Grantchester, this became impossible. The last train to their little village, which they could barely catch if they saw a play, left London at midnight. And there were still eggs to be turned and lamps to be adjusted when they got home to Merton House, no matter how late the hour.

One of Beatrice's most distasteful assignments was making entries in the "Dead Book." This attended the opening of unhatched eggs, which was done on a morbid assembly line. With Bateson and Punnett, Beatrice would retire to the outbuilding that housed the incubators. She sat at a table with a notebook, her husband on one side with a large bowl and a blunt-bladed knife before him, Punnett on the other side gripping a pair of scissors. Bateson would pick up an egg and read off the numbers of the pen, the hen, and the date of laying. "Have you got that, Beatrice?" he would ask, then would stab the egg and peel the shell into his bowl, narrating the oddities of the embryo thus revealed. "Light down, no colored ticks visible, rose comb, no extra toes, feathering on leg," he would recite, and his wife would dutifully record "lt., nts., r.c., n.e., f.l." Then Punnett took the embryo and slit it open to reveal the sex glands — he would call out "male" or "female," and Beatrice would write it down. The two men often cast good-natured bets about which sex the chick would turn out to be. "Altogether it was a messy job, and 'openings' were not much looked forward to," the unflappable Punnett later recalled. But his good nature about such tasks explains why he and Bateson, who was nothing if not overbearing, got along so well.

With the crowd growing impatient, Bateson was ready to deliver his address to wrap up the day's proceedings at the British Association meeting of August 19, 1904. We do not know exactly what he said — the account in the next month's *Nature* was brief and paraphrased, and Punnett recalled only that his words were "striking" — but we can picture him saying it: tall, bearish, deep-voiced, with a flamboyant style of speech that made his every pronouncement sound as weighty as the Sermon on the Mount. No doubt he repeated some of the criticisms of the biometricians that he had been making for years, such as their erroneous belief that Mendelism was disproved because there were so many exceptions to its basic laws. "Arguments built on exceptions" only reveal the paucity of one's own evidence, Bateson might have said. By focusing, for instance, on the "fluctuation and diversity in regard to dominance," he might have accused Weldon of "merely indicating the point at which his own misconceptions begin."

"Soon every science that deals with animals and plants will be teeming with discovery, made possible by Mendel's work," Bateson had said in other speeches. "Each conception of life in which heredity bears a part — and which of them is exempt? — must change before the coming rush of facts."

Bateson concluded with a flourish and returned to his seat. He and his associates were sure that it was all over now, that they had delivered the final blows to the biometricians. When Karl Pearson, Weldon's great friend and defender, rose from the audience, the Mendelians felt especially victorious. Pearson proposed a three-year truce. Why would he do so if he had not felt that he and Weldon were about to lose?

Yes, a truce could be a good idea, agreed the meeting's chairman, the mild Reverend T. R. Stebbing, a self-described "man of peace." It was Stebbing's job, according to Punnett's memoir, to wrap up the day's proceedings. Yes, compromise is a good thing, he said. Mediation is useful. High feelings only cause trouble. The crowd fidgeted in irritation. Was this to be the tame and boring end to such a rousing afternoon?

"You have all heard what Professor Pearson has suggested," Stebbing said. Then he paused, looked around, took a breath. The crowd fidgeted some more. "But what I say," he went on, his voice suddenly loud and forceful, "is: let them fight it out!"

And fight it out they did.

17

A Death in Oxford

I was determined to know beans.

— *Walden,* Henry David Thoreau, 1817–1862

RAPHAEL WELDON SEEMED to gain new fervor after the British Association meeting, though he was mortified by his student Darbishire's defection, his enemy Bateson's eloquence, and what he saw as his own overwrought speechifying. Late the following year, in the autumn of 1905, he saw his chance to retaliate.

The opportunity came in the form of a paper submitted to the Royal Society by one of Bateson's most visible supporters, Colonel C. C. Hurst. Weldon, as chairman of the Zoological Committee, was one of the first to see it, and the paper made him inexplicably furious — setting him on a course of action that would prove to be fatal.

Hurst had come up with a theory about the transmission of coat color in horses based on data recorded in Weatherby's twenty-volume *General Stud Book of Race Horses,* the bible of the Ascot set. After analyzing pedigrees from the *Stud Book,* he concluded that bay and brown were passed on as simple Mendelian dominants, and chestnut as a simple Mendelian recessive. He asked Bateson, as a member of the Royal Society, to sponsor his paper on horses for publication in the society's *Proceedings.*

Bateson was always a little uncomfortable with Hurst's ideas; he believed, as Punnett put it, that Hurst was "over-apt to find the three-to-one ratio in everything he touched." But despite his discomfort, he submitted Hurst's paper. And when Weldon read it, he threw himself into trying to prove Hurst — and, by extension, Bateson — wrong about the distribution of bay, brown, and chestnut coats among horses.

At first Weldon had trouble tracking down the complete *Stud Book*. He found the last four volumes in the Bodleian Library ("The Bodley") at Oxford, where he had been a professor since 1900. But no matter how he rifled through card catalogues — looking for the book under the headings Weatherby, Jockey Club, Horses, Race Horses, Racing, Studbooks, Turf, Sport, and Race — he could not find the other sixteen volumes.

"For a whole day I raged," Weldon said, "and came back despairing. Next day I raged worse."

At last Weldon found a librarian who knew something. "Oh, yes," he said, smiling. "*The General Stud Book* is entered under General, of course." Weldon, infuriated by that "of course," spat back, "Why not under The?"

All twenty volumes finally in hand, Weldon spent every day for the next month looking for inconsistencies. He worked on it seven hours a day, then eight hours a day, then nine. He found a few crosses that proved to him that chestnut was not always transmitted as a recessive. When two chestnuts gave birth to a bay or brown foal, it showed that two supposedly recessive parents could create a dominant offspring — which was, according to the Mendelian law of dominance, impossible.

On December 7, 1905, Hurst presented his paper to the Royal Society. Afterward Weldon rose and grandly announced the anomalies he had uncovered.

Those are just errors of entry, Hurst casually replied, just slips of the pen or the eye. After all, bay (dominant) and chestnut (recessive) are both reddish brown and look the same in a certain light. But Bateson was less sanguine about these anomalies than was

Hurst; he relinquished his sponsorship of the paper and withdrew it from consideration for publication in the *Proceedings*. Weldon took his seat, feeling triumphant.

Within a few weeks, however, Hurst had collected evidence proving that most of the incongruities Weldon had found were indeed due to errors of entry. The mistakes occurred most often with stillborn foals, whose actual coat color did not matter to their owners; they cared about physical traits only in animals that could eventually earn them money. And then there was the case of Ben Battle, a thoroughbred who was registered in the *Stud Book* as a chestnut. Ben Battle, when mated to chestnut mares, had sired bay horses; how could this be if the chestnut parents were both double recessives?

The answer was found in an obscure racing manual called *Form at a Glance*. "Ben Battle never ran as a chestnut," Hurst announced after consulting the manual in early 1906. Ben Battle was a bay, not a chestnut, so naturally his offspring were bays. They were F_1 hybrids — bay father, chestnut mother — whose coat color revealed the father's dominant trait.

With this new information, Bateson offered to resubmit Hurst's paper, which now contained a footnote explaining the apparent inconsistencies. Something about that footnote drove Weldon nearly mad with fury. He used "stronger language than I have ever heard him use" to rage against the "tone and contents" of the footnote, according to his friend Pearson. And he threw himself into an even more impassioned search for contradictory evidence.

This time he was consumed for months. In late winter, "overwrought and overworked," in Pearson's words, Weldon took a long-delayed vacation in Italy. But he took the *Stud Book* with him. "I really want a holiday," he said, "but I cannot leave this thing unsettled." He did no sightseeing, took no rides in the countryside, enjoyed no lavish Italian meals. His letters from Rome were about nothing but horses, obsessed as he was with the ancestry of brown and bay and chestnut thoroughbreds. On his return to Oxford, he pursued his research in stables nearby, searching for a chestnut mare that, when crossed with another chestnut, had given birth to a brown or bay foal. During Easter break at the beginning of April 1906, the Weldons and the Pearsons went together to a little inn at Woolstone, at the foot of White Horse Hill. Once again the holiday was no holiday; Weldon spent the whole time poring over the *Stud Book* and talking to Pearson about the paper he was writing on the inheritance of coat color in horses.

On Palm Sunday, April 8, Weldon cycled into town to develop the photographs he had taken of White Horse Hill. He met Pearson along the way and the two men stopped by the roadside for a cigarette. The bike ride, Weldon admitted, had tired him out, a rare confession for him. However, he took a long walk on Monday and got home late. He was tired Monday night, tired still Tuesday morning, and after breakfast he returned to his bed, where Pearson found him in the afternoon. It looked to be an attack of influenza.

On Wednesday, April 11, Weldon had had enough of bed rest, and over his wife Florence's protests — she saw that he was still weak and coughing — he went into Oxford to visit a picture gallery. The following day he kept an appointment with his dentist. But by then his stubborn refusal to admit that he was ill finally broke down. He went directly from the dentist's chair to the doctor, and straight from the doctor to the hospital. He might not have wanted to stay in bed, but by then he was too sick to get up.

When Florence Weldon received a telegram dispatching her to the hospital on Thursday afternoon, she was hardly surprised. She had known her husband was ill. She had begged him not to venture out every morning since Monday, when it was clear he was suffering from influenza or some other serious infection. Now she rushed to Raphael's side. His naturally gaunt face was deathly pale, his dark hair plastered to his forehead. He could barely catch his breath. All those cigarettes, all that nonsense with the horses, all that pushing himself to the very limits of what his slim, frail body could do, had come to this.

Despite Weldon's popularity in the university town of Oxford, where he and his wife had lived for six years, Florence's vigil was a lonely one. The couple had never had children. Their friends were away for the Easter holidays, enjoying the emerging spring, as were Weldon's students and collaborators. When he died the following day — it was Good Friday, Friday the thirteenth — the forty-six-year-old Weldon had a death as close to a "good" one as is possible: quick and painless, while he was in the full flower of his intellectual vigor.

"So passed away . . . a man of unusual personality," said Pearson, who had last seen Weldon three days before he died, "one of the most inspiring and loveable of teachers, the least self-regarding and the most helpful of friends, and the most generous of opponents."

That last was meant, most likely, as a slap at William Bateson, whom Pearson considered a distinctly *un*generous opponent. Soon a rumor was circulating that Bateson had killed Weldon — not the first such rumor about a man whose bearing and temperament did occasionally make him seem murderous. In Bateson's early years on the staff at St. John's College, legend had it that during his research trip to the Russian steppes, Bateson had shot a man.

We do not know much about the fate of the anonymous Russian, but as for Weldon, what killed him was not William Bateson but overwork. "I cannot leave this thing unsettled," he had said about the *Stud Book* controversy — though why he took it so to heart was never clear. Even when he showed the first signs of illness he refused to put aside his work or come in from the dank, chilly stables. Bateson did not kill Weldon; his own obsession did.

Bateson felt the news of Weldon's death as "a shock — in the literal sense." In its aftermath he fondly recalled the early days of their friendship, fretted over whether he should write a letter of condolence to Florence (would she take it as a greater offense if he wrote to her or if he failed to?), and wondered how he and his associates would fill the vacuum left by Weldon's death. "Now suddenly to have one of the chief preoccupations of one's mind withdrawn, leaves one rather 'in irons,'" he told Beatrice, using sailors' slang for a vessel that fails to come about or fill away — in other words, becalmed against his will, with no wind to fill his metaphoric sails.

Bateson's partner Punnett felt similarly marooned. While the constant rancor might have been demoralizing at times, its absence was in a way even worse. "It kept us on our toes," Punnett said about the fireworks that punctuated his first three years with Bateson, "and added a spice to the work."

Now nothing remained but the work itself.

The work, though, had changed. No longer was it so important to amass bits of evidence to support the Mendelian view of discontinuous variation or to establish that the monk's laws could reliably explain heredity. Scientists who tried, after the rediscovery, to replicate Mendel's work had found that — with certain refinements made possible by twentieth-century advances — his surprisingly modern approach to data analysis had laid an excellent foundation for an emerging branch of knowledge.

By the time of Weldon's death, then, experimental confirmation of Mendel's laws was becoming secondary to a more important mission: creating the new field of study known as genetics, and lining up the best scientific minds of the day to declare themselves on the side of the Mendelians.

18

Inventing Mendelism

> *Gardening has compensations out of all proportion*
> *to its goals. It is creation in the pure sense.*
>
> — "Against Gardens," Phyllis McGinley, 1905–1978

WILLIAM BATESON UNDERSTOOD the power of language. He read
Balzac and Voltaire in the original and considered any man a
Philistine who could not converse easily in French. He subscribed
to German-language newspapers to keep up his colloquial skills,
and corresponded with his German colleagues — Erwin Baur of
Berlin, Hans Przibram of Vienna, Gregor Mendel's nephew Ferdinand Schindler in Moravia — in their own language. He mounted
an aggressive campaign in favor of studying Greek when the ad-
ministration at Cambridge University considered dropping it as an
admission requirement. And he was poised now to turn another
kind of language — the language of science — toward his own ends
by creating a universal terminology that could be used and understood by scientists everywhere.

"Genetics" was the word with which he hoped to begin. It came
from the Greek *genētikos,* meaning "origin" or "fertile" or "productive." Although it later fit euphoniously with "gene," that word was
not coined until several years later. Indeed, Bateson knew nothing
of the word, and in the first decade of the century he had little understanding of the concept of the gene.

He invented the word "genetics" in early 1905, when he was

asked to devise a plan for a new "Institute for the Study of Heredity and Variation" at Cambridge. Call it instead an "Institute of Genetics," Bateson suggested, and endow a "Chair in Genetics" for the man who would direct it. Bateson, of course, was hoping to be made director himself. After twenty years at Cambridge, he was still without a regular academic appointment, squeezing out a living from temporary fellowships and lectureships and a job as house steward of St. John's College. This professorship would be perfect for him; who else in all of Britain was more closely aligned with this burgeoning field?

But the "Institute for the Study of Heredity and Variation" never materialized, nor did the "Chair in Genetics." The money that would have gone to Bateson was used instead to endow a professorship in protozoology.

In 1906 Bateson took his new word to the Royal Horticultural Society, sponsor of the International Conferences on Hybridization and Plant Breeding. At the first of these meetings, held in London in 1899, Bateson had met De Vries and given his prescient, pre-Mendelian talk about arriving at some laws of inheritance by statistically analyzing data over many generations. The second international meeting, in 1902, had taken place at the Brooklyn Institute and was the occasion for Bateson's first journey to New York City. He was stunned by the telephones, the typewriters, the black squirrels in Central Park, the heat of the summer and the warmth of his reception. "At the train yesterday," he wrote to Beatrice, "many of the party arrived with [my book] 'Mendel's Principles' in their hands! It has been 'Mendel, Mendel all the way', and I think a boom is beginning at last." He spent a breathless ten days in New York, thrilled by the metropolitan tempo but convinced that "if I lived [at] this pace for a fortnight I should be like milk that has passed [through] a separator."

In July 1906 he was back on more familiar ground in London, at the Third International Conference on Hybridization and Plant

Breeding. With Bateson at the helm, it was almost a reprise of the British Association for the Advancement of Science meeting in Cambridge two years earlier, when he had trounced the biometricians. The difference — besides the fact that Weldon was dead — was that this time Bateson's hidden agenda was grander than in 1904. He was intent on creating a new science.

In his opening address, Bateson made his case for genetics. The word, he said, "sufficiently indicates that our labors are devoted to the elucidation of the phenomena of heredity and variation: in other words, the physiology of Descent." Bateson knew that if the conferees adopted his terminology, it would forever change the nature of the profession, turning it away from practical plant breeding toward the theoretical and scientific underpinnings of inheritance and evolution.

All right, said the hybridizers and horticulturists in the audience; we'll call the field genetics. We'll agree, starting now, to publish our proceedings and papers under that new rubric. This consensus, which was what Bateson had hoped for, had one unintended effect: it led to an odd rewriting of history. The 1906 meeting was known forever after, in conversation and in print, as the Third International Conference on Genetics — even though there had never been a First International Conference on Genetics, nor a Second.

Now other pieces of the new lexicon started to fall into place. Years before anyone had words to express the difference between genotype and phenotype, years before they even had the word "gene," years before anyone really understood what genes or genotypes were, scientists working in the new field of genetics were linked by their vocabulary. Bateson had already introduced four other new terms, and now they were resurrected: "zygote" — from the Greek *zugōtos,* meaning "yoked" — to describe the fertilized egg, the organism formed by the yoking together of two gametes; "homozygote" (from the Greek *homos,* meaning "the same") and "heterozygote" (from *heteros,* meaning "different") to describe, respectively, purebreds and hybrids; and "allelomorph" (a compound of *allēlēn,* meaning "of one another," plus *morphē,* meaning "form") to describe the different versions of a particular trait. Tall and dwarf, for instance, could be thought of as two allelomorphs for height in Mendel's pea plants. Later allelomorph was shortened to allele, the word still used to describe the many normal variations possible for a single gene.

Scientists of the day also accepted Bateson's terminology for many of Mendel's most confusing concepts. To differentiate more clearly among the generations in an experiment — which Mendel called "first hybrid generation," "second hybrid generation," and so on — Bateson introduced a new notation: P1 for the parents (and P2 for grandparents, P3 for great-grandparents) and the letter F, for "filial," to represent the offspring: F1 for the children, F2 the grandchildren, F3 the great-grandchildren.

This universal language was the first step in turning the emerging science of genetics into a coherent discipline. The second step was to give it an emotional center. That is where Gregor Mendel came in. Every new science needs a hero — someone on whose giant shoulders his disciples can stand — and Mendel was an easy man to lionize. Partly because so little was known about his actual biography, and because the little that was known was so admirable — solitary, devout, gentle, humorous, modest — the Moravian monk was an ideal *tabula rasa* on which latter-day Mendelians could etch a tale that it pleased them to think was true. He also was a good father figure because his approach to data collection and analysis was so thoroughly modern. Mendel was a twentieth-century scientist trapped in the nineteenth century, and the romance of his story was that he died unappreciated and embittered, mired in a dreadful silence that engulfed him and his reputation for thirty-five years.

But none of this would have caused anyone to turn Mendel into the founding father of genetics if it were not for the much more critical point: Mendel had been right.

The mythologizing that turned Mendel from monk to hero reflected a process that is central to the formation of almost any new science. Most researchers, even in Bateson's day, spend most of their time in laboratories that look and feel like factories. If they can keep in mind some brilliant founding father, some scorned or unappreciated genius in whose footsteps they now loyally tread, they can more easily maintain their sense of mission through even the most routine, inconsequential chores.

Over the next few years Bateson hit his stride. In 1907 he was invited to America to give the prestigious Silliman Lectures at Yale University in New Haven, Connecticut. He also attended the International Zoological Congress in Boston and visited his fellow "geneticists" — the word was not yet in common use on either side of the Atlantic — in Woods Hole, Massachusetts; Cold Spring Harbor, New York; Manhattan; and Toronto. He was abroad from July until November. His welcome was so effusive that sometimes he had to force himself to keep a level head. "The Americans are rather absurd in their hero-worship and one has continually to remember that they keep a constant procession of heroes on the march," he wrote to Beatrice in late August as he was en route from Massachusetts to New York. "But after the years of snubbing it is rather pleasant to get appreciation even though in an overdose. I go about like a queen-bee on a comb, and I should not be surprised to see some admirer backing out of my presence."

It was during this long sojourn that Bateson met Thomas Hunt Morgan of Columbia University, already the leading genetics researcher in the United States. For Bateson, Morgan was the reincarnation of Raphael Weldon — not because of any physical resemblance but because of something in his manner and his scientific ideas. Bateson turned against him almost by instinct.

At the time of their first meeting, Morgan was engaged in research that he had entered into much the way Weldon had approached his last, fatal encounter with the horses. Like Weldon, he was trying to demonstrate that a Mendelian interpretation of a series of animal experiments was wrong. The object of Morgan's fierce attention was Lucien Cuénot of Nancy, France. In 1905 Cuénot found that coat color in mice seemed to involve three, rather than two, allelomorphs: yellow, which was dominant; gray, which geneticists called agouti; and black. What was most surprising was that when he crossed the dominant yellow mice with either gray or black, then crossed the F1 hybrids with each other, his results in the F2 generation were distinctly non-Mendelian: two yellow hybrid mice for every nonyellow (gray or black) recessive, with not a single pure-breeding, double-dominant yellow to be found. Mendel's 1:2:1 ratio had been reduced to a ratio of 2:1.

Cuénot's reaction was to reassess Mendel's assumption that all possible combinations of gametes have an equal chance of occurring. He proposed instead the notion of "selective fertilization," in which the union of two particular gametes — in this case, yellow with yellow — just does not work, almost as though like were repelling like. Because he was basically a Mendelian, Cuénot saw his "selective fertilization" hypothesis as simply embellishing Mendel's original findings, not refuting them. But Morgan, who at the time was an outspoken anti-Mendelian, thought Cuénot's 2:1 ratio was disproof of the central tenet of Mendelism — the law of segregation, which Bateson was calling "the purity of the gametes." To Morgan it was evidence instead of what he had been saying all along: "once crossed, always mixed." There were no pure-breeding yellows, said Morgan, because all of them had been infected, in the F1 generation, by mixing with nonyellow allelomorphs in the hybrid state.

The only reason Mendel failed to see any latent effect on his peas of having once been mixed as hybrids was that he did not carry through his tests for enough generations, Morgan said. In yellow mice, he assumed, hybrid contamination must simply show up more quickly than it does in yellow peas.

Bateson and Morgan quickly became adversaries. Morgan had publicly questioned Mendel's laws — and Bateson was primed for a fight. He found the prospect of a new controversy strangely satisfying, as though he needed a good hearty struggle to enliven his suddenly respectable middle age.

So much about Thomas Hunt Morgan was the antithesis of Bateson. Born the same year that Mendel's *Pisum* paper was published, he was five years younger than Bateson, which can be a significant span for men in their forties. In 1907 Morgan was a relatively youthful forty-one; Bateson, at forty-six, was undeniably aging and still without a professorship or even a reliable income. Morgan was raised in the rolling bluegrass countryside of Lexington, Kentucky, which gave him a drawly southern aura that to Bateson, ever the intellectual snob, seemed "rough" and uncultured. "TH Morgan is a thickhead" he wrote once to Beatrice. Even fourteen years later — when their conflicts had been resolved and they met for a last time in New York — Bateson still found Morgan was "of no considerable account," with a range of interests and abilities that was "dreadfully small." He was bothered by his feelings, however — he was, after all, a guest during that visit in the Morgans' home — and he wailed to his wife, "I wish I liked Morgan better."

One thing that did unite the two men was a common wrongheadedness. In spite of their many differences, both resisted the emerging "chromosome theory" of inheritance. Chromosomes had been spotted under the microscope by several scientists, including Karl von Nägeli, in the 1840s, but none of them quite understood what he was seeing until the German anatomist W. von Waldeyer gave them their name in 1888. He called them chromosomes because they darkened when treated with chromatin, a stain that scientists used to see things more easily under a microscope. Even then, no one knew for sure where chromosomes usually resided or what they did, since they were clearly visible only during the brief reproductive phase of a cell's life cycle, not in the normal resting state. Much of what was known about chromosomes in the late nineteenth century was based on inference, so no one could even say with confidence whether the chromosomes found in the daughter cells, after cell division, were the same as the ones that had been in the precursor cells, or whether they arose entirely *de novo* every time a cell split in two.

A belief that the chromosomes contained some important inheritable material, then, depended first on a conviction that the chromosomes reassembled themselves after each cycle of cell division — a belief, in other words, that they were eternal. Walter Sutton of Columbia, a graduate student in the early 1900s, was able to recognize certain chromosomes under his microscope before cells divided and could show that these same chromosomes, with the same characteristics, appeared each time a cell split to form two new ones. Cell splitting was known as mitosis (from the Greek word *mitos,* for "thread," describing the appearance of the chromosomes right before the cell divides). He also saw that the same chromosomes reappeared after reduction division, or meiosis (the Greek for "diminution"), the cell-splitting process that creates cells with only half the number of chromosomes, which eventually become the male or female gametes.

Sutton's conclusion was that the same chromosomes that lined up in matching pairs at the moment the zygote was formed persisted throughout an organism's lifespan. Through every cell splitting, through every reduction division, the chromosomes reproduced themselves, retaining their identity through countless cycles of mitosis and meiosis. "The association of paternal and maternal chromosomes in pairs," he concluded in 1902, "and their subsequent separation during the reducing division . . . may constitute the physical basis of the Mendelian law of heredity." This was the first expression of what would come to be known as the chromosome theory of inheritance.

At about the same time that Sutton was working in New York, Theodor Boveri, a professor of zoology and comparative anatomy at the University of Würzburg, in Germany, was drawing similar conclusions about the function of the chromosomes. Boveri showed that the number of chromosome pairs is fixed for each living organism. He found four in the roundworm; twelve in the mole cricket; sixteen in the cat, wheat, and birch; twenty-four in the edible snail, salamander, lily, tomato, and man; thirty-two in the earthworm; thirty-six in the electric ray. He was not right about all of the specifics — man, for instance, has twenty-three pairs of chromosomes, not twenty-four, and the cat has nineteen rather than sixteen — but he was right about the concept of constancy within a species and the idea that the total number of chromosomes is significant.

Further proof came from Boveri's work with sea urchins. He used multiple fertilization and other laboratory tricks to create embryos with unnatural numbers of chromosomes and found that the only ones that developed into viable sea urchins, with all of their body parts in place, were those with the correct chromosome complement of thirty-six. He thus showed that the sea urchin, like every other species he examined, ordinarily carried a predetermined number of chromosomes, and that any variation in that number resulted in large-scale developmental abnormalities. He tried, but failed, to demarcate which of the chromosome's "different qualities" — his term for the still-unnamed genes — accounted for which particular physical traits in the animal.

Boveri's publications about the invariability of chromosomal numbers in each species were perfect complements to Sutton's publications about the same time, concerning the permanence of specific chromosomes within specific cells. Taken together, the findings indicated that important hereditary information resided on the chromosomes — information that scientists were just starting to believe was carried, as Mendel's paper had suggested, as "units" or "factors." There was still some resistance to full acceptance of this idea, based largely on its discomfiting similarity to the theory of preformationism. What was the difference, really, between believing an organism capable of carrying within itself all its fully formed descendants in one microscopic packet, and believing it capable of carrying coded information about those descendants in a few invisible strands inside the nucleus?

These objections were generally stilled after enough evidence accumulated to support the hereditary significance of the chromosomes. By 1904 Walter Sutton and Theodor Boveri, an American and a German who never met, found their names linked in one of the most dramatic ideas of early twentieth-century biology: "The Sutton-Boveri Chromosome Theory of Inheritance."

Bateson and Morgan would have none of it — though for different reasons. Bateson was offended by the mechanistic assumptions of the chromosome theory. He preferred a more holistic approach to inheritance, which brought him to such notions as the "presence and absence theory," the "theory of repetitive parts," and, his special favorite, the "vibratory theory." That last would become "a common-place of Education, like the Multiplication table or Shakespere, before long!" he wrote to his sister in 1891, just days after the vibratory theory had occurred to him in a burst of inspiration. Even after becoming Mendel's bulldog, Bateson never completely abandoned his own pet idea, which related to waves, sand patterns, ripple marks, and zebra stripes, all patterns that repeat and undulate in nature.

Morgan, for his part, had by 1904 come around to accepting much of Mendelism, in particular its presumption of discrete units to account for the inheritance of particular character traits. But he believed that the chromosome theory actually contradicted the essentials of Mendelism. If each chromosome carries many determinants — as each one must, if all of an organism's traits are to be accounted for — then why do so few determinants seem to travel in tandem as they move from parent to offspring? "Since the number of chromosomes is relatively small and the characters of the individual are very numerous," he reasoned, "it follows on the theory that many characters must be contained in the same chromosome. Consequently, many characters must mendelize together. Do the facts conform to this requisite of the hypothesis? It seems to me that they do not."

A stubborn refusal to accept the chromosome theory seemed for a while to unite these two men. But their stubbornness led to two opposite results. For Morgan, who embarked on more determined investigations to come up with an alternative to the chromosome theory, it led to an epiphany that enabled him to reconsider his blind spot and go on to contribute some of the most significant findings in the field — all related to the function of the chromosome in relation to the gene and heredity. For Bateson, who dug in his heels and worked hard to come up with alternative explanations, even as evidence for the chromosome theory continued to mount, it led instead to his being relegated to the back lots of scientific history, as the field he had named spun past him and he steadfastly refused to admit that, back in the beginning, he had been wrong.

Unlike Bateson, Morgan made a habit of admitting to his earlier mistakes. Regarding his attempt to reinterpret Cuénot's observations with the yellow mice, and thereby to turn Mendelism upside down, Morgan found that his alternative explanation just did not work. He tried to test his hypothesis on other strains of mice, but none of them showed a reappearance of dominant traits in the recessive strain, as his "once crossed, always mixed" dictum would have anticipated. "It is evident that the hypothesis failed when tested," he concluded, "and must therefore be abandoned."

At around this time, an American geneticist offered a third hypothesis regarding Cuénot's findings. William Castle of Harvard suggested that carrying two yellow allelomorphs was somehow lethal to these mice before birth. This would explain why the first piece of the Mendelian ratio of 1:2:1 — the "one" that referred to double dominants — was missing, leaving only a 2:1 ratio for Cuénot to observe and try to explain. Other biologists, such as Bateson's friend Erwin Baur of Berlin, were also uncovering anomalies — in Baur's case, in snapdragons — caused by the failure of a single class of zygotes to survive. In other words, according to Castle, Mendel's law was right, but the specifics of these mice or snapdragons led to a confusing anomaly. This explanation was soon confirmed by other investigators, who dissected the dead embryos of Cuénot's strain of yellow mice and found that all were double dominants.

Such *post hoc* explanations for apparently contradictory results checker the history of genetics, as they do in just about any emerging science. Only much later, after investigators have gone off in entirely wrong directions — or, in the case of Mendel, been flummoxed and given up entirely — do scientists understand what had seemed to be errors in either research design or underlying theory: Mendel's failure to replicate his results in *Hieracium*; Correns's findings of unexpected numbers of colorless *Pisum*; Cuenot's missing yellow mice. Bateson encountered a similar stumbling block in 1899, when he crossed spiny and smooth varieties of *Datura* and expected all his F1 generation of hybrids to be spiny. He found, instead, that a significant minority had inexplicable smooth patches. It was not until nearly twenty years later that the "patchiness" was shown to be a symptom of infection with a previously unidentified plant virus known as quernica. The moral here, if there is one, is that sometimes it makes sense to hold tight to a cherished theory, even in the face of some apparent inconsistencies.

Eventually Bateson came around to accepting the chromosome theory of inheritance. But it was late in the game — just two years before his death — and historians still wonder how genuine his conversion really was. "I don't like it," he grumbled about his change of heart, "but I see no way of escape."

In 1909 the new field of genetics turned a corner with the introduction of the word "gene." An array of earlier words to represent the concept of a discrete heritable factor had been tried and rejected: physiological units, gemmules, micella groups, pangens, plasomes, idioblasts, biophores. None of these terms stuck, partly because each was intimately connected with a theory of inheritance that ultimately fell apart. That was, perhaps, the brilliance of "gene."

Wilhelm Johannsen, a professor of plant physiology at Copenhagen Agricultural College in Denmark, coined the word in 1909, four years after Bateson introduced "genetics." He derived it, he said, not as a contraction of genetics, but as a nod to Darwin's theory of pangenesis, from which De Vries had, in turn, created the word "pangen." Johannsen said he simply cut off the first syllable of De Vries's word and turned the second syllable into something entirely new. "The word *gene* is completely free from any hypotheses," he said. All it indicates is that "many characteristics of the organism are specified in the gametes by means of special conditions, foundations, and determiners which are present in unique, separate, and thereby independent ways — in short, precisely what we wish to call *genes*."

Along with "gene," Johannsen introduced two other words that proved to be just as central to the emerging lexicon: "phenotype," meaning an organism's appearance; and "genotype," meaning its genetic makeup. This was a distinction Mendel had intuited more than fifty years earlier, before anyone had the ability — either conceptually or linguistically — to name it. In the case of an organism showing a recessive trait, genotype could be inferred from phenotype, since the only way a recessive trait showed up in the phenotype was when the organism was double recessive. But when dominant traits appeared, further experimentation was required to see what the genotype was. The organism could be either a double dominant or a hybrid.

Johannsen's approach to the word "gene" was almost exactly the opposite of Bateson's approach to "genetics." He staked no personal claim on it, made no effort to direct its use. He allowed it to change, chameleonlike, to suit whatever theory of inheritance was being explored at the time. While both words have stuck, the notion of the gene is the more basic and the more necessary. Without a word for the unique determiners of heredity, geneticists of the early twentieth century would have been unable to talk to each other, rendered as mute as two far-flung Indian tribes who can communicate only through the vague, evanescent spurts of smoke signals.

With the new word to guide him, and his new conviction that the chromosomes were essential players in the drama of inheritance, Thomas Hunt Morgan set to work looking for specifics.

He began, much as Mendel had, by focusing on a single organism. For Morgan this was an insect known colloquially as the fruit fly, though *Drosophila melanogaster* is more properly called the vinegar fly because of its preference for overripe fruit with some tang to it. Morgan studied *Drosophila* chiefly out of expediency — combined, as in Mendel's case, with a little bit of luck. His cramped laboratory at Columbia, less than four hundred feet square, was already stuffed with live pigeons, chickens, starfish, yellow mice, and rats. So when a new student wanted an organism to work on, there was no room for anything much bigger than a fly.

Drosophila were cheap to feed and house; all they needed were a few ripe bananas and some bottles for breeding. Morgan bought the bananas, but he and his lab assistants swiped empty milk bottles from the stoops of upper Manhattan on their morning walks to work. To this day, following a tradition begun in Morgan's "fly room" at Columbia, many scientists who work with *Drosophila* breed them in milk bottles — although now they usually pay for them.

It turned out that vinegar flies have one characteristic that makes them invaluable to geneticists: nice big chromosomes. They have just four pairs altogether, easily visible under even the low-powered microscopes of Morgan's day. In addition, they reach sexual maturity within a week, and females bear several hundred baby flies at a time. *Drosophila* is the geneticist's pet because it breeds constantly, prolifically, euphorically.

No matter how many offspring are created, however, they tend to all look the same, which is as bad for a geneticist as a windless sea is for a sailor. For two years Morgan subjected thousands of vinegar flies to chemicals, toxins, and x-rays in hopes of inducing some interesting change. He failed to find a single one. He was looking for mutations in part because of Hugo De Vries's theory, which held that species arise by mutations "suddenly, . . . without any visible preparation and without transition." Morgan had visited De Vries in Holland and was drawn to the Dutch botanist's emerging skepticism about Mendel's laws — a skepticism that matched Morgan's own. De Vries was hoping to eclipse the rise of Mendelism with his own mutation theory, according to which either a single character or a whole set of characters can change without warning through some alteration in the makeup of its determinants. De Vries did not really try to explain exactly how this would happen, other than to say that species cycle through periods of "mutability" and "immutability."

Finally, in the early days of 1910, something interesting turned up under Morgan's microscope. Among a field of normal red-eyed vinegar flies, there appeared a single fly whose eyes were white.

Morgan knew this fly was precious. He brought it home in a jar each evening, keeping it next to his bed while his wife, Lilian, was in the hospital having a baby. According to family lore, Morgan went to visit her in the hospital during this time. "Well, how is the white-eyed fly?" Lilian asked. Morgan spent quite a long time explaining the fly's precarious condition before remembering to ask, "And how is the baby?"

The following week the white-eyed male was ready to breed, and Morgan crossbred him with red-eyed females. He got 1,237 hybrids from this cross — all with red eyes. This was consistent with Mendel's law of dominance: the recessive trait (white eyes) seemed to disappear in the first generation of hybrids. When crossed with each other, these hybrids gave rise to an F2 generation that also confirmed Mendel, this time his law of segregation. In the F2 generation about three-quarters of the flies had red eyes and one-quarter had white: a Mendelian ratio of roughly 3:1.

These findings flew in the face of Morgan's "once crossed, always mixed" dictum. The two eye colors had been mixed in the hybrid F1 generation, but here they were, separating out in the hybrid's offspring, totally unmixed — not a pink eye in the bunch.

More perplexing was what became of this 3:1 ratio if you recalculated it according to sex. Among the males, half had the normal red eyes and half the abnormal white. Among the females, all had red eyes. Put another way, all the white-eyed *Drosophila* were males.

In homage to De Vries's mutation theory, Morgan called the white-eyed trait a mutation. But this similarity in vocabulary belied a difference in intention. To De Vries a mutation was interesting only because of what it told him about evolution. Indeed, he believed that a mutation, which he defined as a drastically changed physical form, was the unit of discontinuous variation, just as the gene was coming to be accepted as the unit of inheritance. To Morgan, on the other hand, a mutation was interesting only for what it told him about the gene. He sought out mutations because of what they revealed about the role of normal allelomorphs in the transmission of characteristics from one generation to the next. The difference in the meanings of the word "mutation" led to a confusion — was a mutation a change in appearance only or also in some essential component of the gene? — that was not cleared up for another twenty years. Only much later, after De Vries and Morgan were both long dead, did Morgan's definition of *genetic* mutation become the universally accepted one.

Once Morgan and his colleagues identified the first white mutation, they began to see mutant flies everywhere. Many of these new mutants revealed a distinctive sex distribution similar to that of the white-eyed flies. It did not take Morgan long to realize that these were "sex-limited" mutations — that is, the gene with the mutation was carried on the sex chromosome. Biologists at the time already knew that females were homozygous for the sex chromosome; they had two copies of the long X chromosome, on which many separate genes could be stretched like, in Morgan's metaphor, "beads on a string." Males were heterozygous for the sex chromosome, carrying one long X paired with a shorter chromosome known as Y. If a female carried the white mutant on one of her X chromosomes, she could carry the normal red-eyed allelomorph on her other X. This would make her hybrid for eye color, with the dominant red being the color that showed. But males could not be hybrid for eye color if the gene for that trait was carried on the X chromosome. If a male carried the white mutant on his single X chromosome, his eyes would always be white, because he did not have another X that might carry a normal allelomorph.

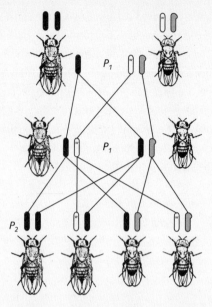

*Morgan's diagram, from his 1926 textbook, of trans-
mission of the white eye mutation in vinegar flies.*

Using this logic, Morgan concluded that whenever a mutation
showed up overwhelmingly in males, it was probably carried on
the X chromosome. This allowed him to calculate how often different mutants arose, how often they were associated with other
mutations in the newly discovered phenomenon called linkage,
and where exactly along the chromosome "string" each genetic
"bead" was located. Within a few years of finding the white-eyed
fly, Morgan and his associates had created a map for the X chromosome — a map that was the precursor for all subsequent genetic
maps, including the gargantuan map now being compiled of the
normal locations of all the 30,000 or so genes that make up the total human complement.

With the discovery and naming of the *white* mutation, another
precedent was set as well: a colorful system of naming *Drosophila*

mutants that persists, almost as a competitive sport, to this day. By calling that first mutation *white*, Morgan created the convention of giving a mutation the name of the abnormality it created, rather than the general trait it represented — in this case, eye color. The second vinegar-fly mutation he and his associates identified was *rudimentary* wings instead of full-sized ones, then a *yellow* body instead of the normal gray. By the end of the century the list of *Drosophila* mutations included words that conjure up vivid images of deformed insects — *ruffled, stumpy, cloven thorax* — and others that seem almost deliberately obtuse: *fruitless, bizarre, coitus interruptus, drop-dead, smellblind, rutabaga, pirouette, stuck, singed, disheveled*. A mutation called *turnip* causes a fly to act just plain stupid, and one called *sevenless* causes it to move toward darkness instead of light. One recently named mutation, *dissatisfaction*, affects mostly females: no male can please a *dissatisfaction* mutant, who, when males approach her to mate, just flicks them all away with her wings.

When Morgan embraced and refined the chromosome theory at the end of the decade, Bateson turned away, all the fight taken out of him. He did not yet accept the theory himself, but he could see that he was becoming more and more isolated in this position. In 1908 he finally received the vote of confidence from the university that he had been longing for: he was made a professor in Cambridge's new department of genetics. But the job was tenuous: it was created by a five-year bequest from an anonymous donor, with no guarantee that the money would be renewed. Bateson still felt that he and his family were living hand to mouth. So two years later, when he was asked to become director of the new and well-endowed John Innes Horticultural Institute, he accepted. For the first time in his life he was leaving Cambridge, starting a new institute from scratch, and relocating his family and his experiments to the John Innes headquarters in Merton Park, a town southwest of London. "A thoroughly new start in middle life is often good for a man," he said, "and I am going to make it." Maybe he was also responding to the old Greek notion of the climacteric, the idea that a person undergoes a significant shift in outlook and circumstance on a regular basis — at the ages calculated by multiplying the number seven by the odd numbers three, five, seven, and nine. In 1910 Bateson was seven times seven: forty-nine.

The year 1910 was significant for another reason as well — one that symbolized the hope and possibilities of a century that was still quite new. In October of that year Bateson traveled to the St. Thomas monastery in Brünn to participate in the unveiling of the Mendel memorial statue. This was the first such international tribute to Mendel, whose work had laid the foundation on which his successors — including Correns, De Vries, Sutton, Boveri, Morgan, and dozens of others involved in the study of cell structure, chromosomal function, and genetics — could build.

The journey, the unveiling, the very existence of the statue itself, were signs that scientists from around the world now agreed that one man, the father of genetics, had pointed them all in the right direction. Now they were traveling together along the path Mendel had marked in pursuit of a better understanding of humankind and our relationship to our own destiny.

19

A Statue in Mendelplatz

The breath of flowers is far sweeter in the air (where it comes and goes, like the warbling of music) than in the hand.

— "Of Gardens," Francis Bacon, 1561–1626

HOW ABSURD FOR THE SPECIAL GUEST from England to be wearing a swallow-tailed coat. Why was he not in an ordinary frock coat at half past ten in the morning? Surely the rules of etiquette in London couldn't be so dramatically different. Here, at least, in the good city of Brünn, in the land then known as Moravia, in the waning years of the Hapsburg Empire, no one — no matter how exalted his social station or how important he fancied himself to be — would be seen in tails until after the sun had set.

But there he was, the larger-than-life biologist William Bateson in his startling formal wear, moving to the podium. The date was Sunday, October 2, 1910. He had just been introduced by Hugo Iltis, a biology teacher at the Brünn Realschule, where Mendel had taught. In honor of his now-famous predecessor, Iltis had assembled some of the world's most illustrious scientists for this signal event. Bateson cleared his throat and began to shout.

Shouting was essential. The ceremony was being held outdoors, and the acoustics were bad. Even by straining his voice, Bateson later learned from his friend Hans Przibram — a professor at the University of Vienna, with whom he stayed en route to Brünn and again on his way home — he could barely be heard in the back rows. Aware that his audience was confused and inattentive, suffering from the stage fright that always plagued him, and embarrassed by the inappropriateness of his attire, which he later described as "ridiculous," Bateson hurried through his remarks. He believed he spoke for no more than four minutes altogether.

How grievous it was, he bellowed, that Gregor Mendel, whose memorial statue was about to be unveiled, had not lived to see his work so grandly revered. How tragic that Mendel, a mild-mannered priest who had spent most of his life as a member of the Augustinian monastery of St. Thomas, had toiled in such obscurity and had died without having any idea how pervasive his influence would be on subsequent generations. But let us not think of Mendel only with sorrow, Bateson intoned. The priest had looked intimately into the face of Truth — and thus had tasted what is surely the greatest pleasure this sorry little world can provide.

Bateson closed his speech with a quote from a 1785 poem by Friedrich von Schiller, one made famous when Beethoven set it to music in 1825 in the magisterial "Ode to Joy" of his Ninth Symphony. *"Alle Menschen werden Brüder,"* Bateson yelled in his rich baritone. This was something everyone heard; his German was easier for the audience to understand, even with the bad acoustics, than his English had been. And the Czech- and German-speakers were great fans of music, especially vocal music. Bateson himself, something of a cultural snob, later wrote to Beatrice that Moravia was "a land of music" and that the two male choruses that sang before the unveiling were "a glory." Such unsolicited praise was rare in Bateson's letters to his wife; more typical were his other comments about the unveiling ceremony, bemoaning the atrocious food and the statue itself, which he called "banal & shocking" in "all its gross absurdity."

Gross absurdity? Shocking and banal? Well, perhaps. The statue, which today presides over the courtyard where Mendel tended his garden peas, was a standard-issue white marble monument featuring a likeness of the man being memorialized, flanked by some items that symbolized his contribution to mankind. It bore, as did so many statues of the day, a face far more Romanesque than its subject's had been. Mendel's face was purely eastern European, broad and flat, "somewhat roughly hewn by nature as sculptor," as one former student put it. Yet for all its inaccurate chiseled handsomeness, something about the statue's face seemed in fact to invoke the human Mendel. It wore the sweet, thoughtful expression that was said — at least according to the legend built around him by 1910 — to have been his typical look. There he was, gazing into the distance as though hoping for an insight into life itself, his arms outstretched as toward the future, surrounded by a bas relief of peas and beans to symbolize his connection to the natural world. At the statue's base, two small kneeling people, also in bas relief, represented sexual love and fertility — the mechanisms, of course, of human heredity.

The effect might seem banal to twenty-first-century eyes, but it is somewhat surprising that Bateson found it so. Perhaps it did not meet the standards and refined sensibilities of a man born and bred in Cambridge, a man who had taught himself, after decades of collecting, to be an art connoisseur.

But it is unclear what Bateson found "shocking" in a memorial that, while ordinary, was above all else inoffensive. Maybe it was because of the botanical errors in the bas relief — the peas climb up a rock, which they would never do in nature, and they have leaves on their peduncles, the main stalks of the flower clusters, where *Pisum* has no leaves. Maybe he was "shocked" by the inscription — not the words, which were bland enough, but the language in which they were written. At the time the issue of language and ethnicity was threatening to split Moravia in two. The German-speakers still had power over the ethnic Czechs, and, as in Mendel's day, they still quashed all attempts at Czech nationalism, afraid that allowing the Czech language to be used would stir up the patriotic fervor of the majority. What a slap it was to the Czechs, then, for the inscription on the monument — "To the Investigator P. Gregor Mendel, 1822–1884, Erected in 1910 by the Friends of Science" — to be in German only. Bateson would have been aware of this friction and, while he probably felt no special allegiance to the Czech patriots, his natural tendency was to side with the underdogs.

What a further offense for the incumbent abbot of the St. Thomas monastery, the Czech national Franciscus Salesius Bařina, to be excluded from the proceedings on stage — an especially glaring omission because he was one of the only men in attendance who had met Mendel, having been admitted to the monastery nearly thirty years earlier by Abbot Mendel himself. And finally, how crass that the little exhibit of Mendel documents, collected and displayed for visiting scientists as part of the celebration, was to be housed not at the monastery, where Bařina would have officiated, but at a local facility known as German House.

The dignitaries who attended the unveiling — as well as those who did not — were, by their very presence or absence, taking a stand about Gregor Mendel and his true role in the establishment of a new field of science, the field that Bateson had dubbed genetics only a few years before. Some men stayed away specifically because they believed too much was being made of Mendel's role in creating this new discipline. These were the precursors of the modern Mendel revisionists, the historians of science who argue that he is lauded as the father of genetics for all the wrong reasons — not because he had (or was capable of) any special insights but because he provided a convenient banner under which the rediscoverers, especially Bateson, could rally troops to win the scientific battles in which they already were embroiled. But this view, then as now, undermines the brilliance of what Mendel *did* accomplish. He might not have known, prior to the twentieth-century understanding of the gene, exactly what his findings meant, but he arrived at them in a logical and sophisticated way, making him decidedly the first in a long line of modern genetic investigators. Using methods quite different from those of his hybridizing peers — but true to his own life story as a monk with time, patience, brilliance, and perseverance — Mendel turned a few good insights into scientific gold. He had the modern sensibility required to carry out his experiments over several generations; to regard each characteristic separately from every other; and to count his results and look for mathematical relationships among them.

True, he might never have understood the laws of inheritance, which by 1910 were becoming clearer because of advances in cell biology, evolutionary thought, and statistical analysis. But it is unfair to allow Mendel's critics, either then or now, to damn him for making only a tentative start in this most oblique, code-driven, and materialistic of the biological sciences. Had he not made that start, however tentative, who can say how genetics would have begun instead — or even when?

Most conspicuous by his absence at the statue unveiling was Hugo De Vries. The Dutch botanist did not want to go down in history simply as one of Mendel's three rediscoverers from the stunning spring of 1900. But to his continuing horror he had spent the previous decade watching his name remain yoked to the name of Gregor Mendel, a predecessor for whom he finally held a feeling approaching contempt. He wanted to be known as an innovative botanist and evolutionist in his own right. He wanted his theories about inheritance and species formation — his refinement of Darwin's theories with his own theories of mutation and of intracellular pangenesis — to be the reasons for his immortality. So he had refused to sign the petition soliciting funds for the marble statue and had pointedly turned down invitations to the event from Hugo Iltis, who would eventually receive acclaim as Mendel's Boswell. Indeed, twelve years later, when the centennial of Mendel's birth provided the occasion for an even grander celebration than the statue unveiling, De Vries was downright rude in his refusal to participate. "The glorification of Mendel," he wrote then, "is a matter of fashion which can be shared by anyone, even without understanding; this fashion is likely to pass."

A second rediscoverer had the opposite attitude. Erich von Tschermak, the Austrian, not only attended the statue unveiling but played a prominent role in the festivities. He relished his identity as a rediscoverer, knowing it would help him reach wider acclaim than his own botanical studies ever would. Tschermak served on the fundraising committee to solicit the 4,000 crowns the statue cost, finally collecting money from geneticists and botanists worldwide, as well as from a few large foundations and from the good people of Brünn. Tschermak gave the second speech of the dedication morning, following the choir performances and a welcoming address by Iltis. This was the kind of grandstanding people had come to expect of Tschermak. A few years after the unveiling, some started questioning whether he should be considered an independent rediscoverer at all, since he never seemed to have understood Mendel's ratios. Later, during the 1930s and 1940s, when Tschermak was an outspoken Nazi and eugenicist, it became easy to dismiss him as a man who did not possess a true spirit of scientific inquiry.

The third rediscoverer, Karl Correns of Germany, did little more for the statue dedication than lend his name to the petition used to solicit funds. He took no official role at the festivities. It was typical of Correns to hang unnoticed in the background — except for that strange day in April 1900, when he had flown into such a rage on finding that Hugo De Vries had beaten him to publication once again. Considering his major role in helping bring Mendel to scientific attention — it was Correns, after all, who named "Mendel's laws" — the fact that he had no official part in the ceremony itself is most decidedly peculiar.

Even a decade after his paper's rediscovery, even in his own home-town, Gregor Mendel in 1910 was not yet the larger-than-life heroic figure he would eventually become. "Who is this Mendel?" one Brünn burgher was overheard asking another, as the two stopped at a bookshop window to read a display requesting contributions for a statue to honor the town's most illustrious citizen. "Don't you know?" asked the second. "He's the fellow who left the town of Brünn a legacy!" This man, so proud of his knowledge, had obviously misunderstood the word *Vererbung*, which has the same two meanings in German as does the word "inheritance" in English: in a scientific sense, it means "heredity"; in a commercial sense, "bequest."

After the fanfare and the dramatic unveiling of the Mendel monument, with its flags and fresh greenery and choirs performing on the Klosterplatz — soon to be renamed Mendelplatz — there remained a vague sense of disappointment. Hugo Iltis, who had worked so hard as local coordinator and official record-keeper, felt a strange sense of anticlimax. Fourteen years later he wrote the first official biography of Mendel — called, in English translation, *Life of Mendel*. In his book Iltis explained that the statue was just a statue, an empty symbol compared to Mendel's true memorial, the science of genetics. "The progress of research since the beginning of the century has built for Mendel a monument more durable and more imposing than any monument of marble or bronze," he wrote in 1924. "Not only has 'mendelism' become the name of a whole vast province of investigation, but all living creatures which follow 'mendelian' laws in the hereditary transmission of their characters are said to 'mendelize.' In these words Mendel's name will be immortalized as long as science endures."

The statue, and especially its location, would take on symbolic significance of its own in the years ahead. In 1950, during the darkest days of the Communist attempts to rewrite the history of genetics, the Czechoslovakian army carted away the statue in the middle of the night. The authorities stripped the stone Mendel of his grand marble pedestal and propped him in a corner of the back courtyard of the monastery, which had rented out the monks' former living quarters to government agencies. No one offered an explanation for the statue's bizarre relocation to such a remote spot. But it was clear, through the events that would occur within the next two decades across the Soviet bloc, that Mendel was no longer considered a hero worth celebrating. This was the era of "Lysenkoism," named after the Soviet agricultural minister Trofim Lysenko, who had risen through the Communist Party ranks by denouncing classical genetics as "heresies." Instead he propounded the belief that plants, animals, and humans could be turned into whatever society needed them to become, irrespective of their genetic limitations, given the proper environment. He was so disdainful of classical geneticists like Gregor Mendel and Thomas Hunt Morgan that one of his first orders as director of the Genetic Institute in Moscow was that all the fruit flies be boiled to death.

The world of genetics changed, even in the Soviet bloc, when James Watson, an American, and Francis Crick and Maurice Wilkins, two Englishmen, were awarded the Nobel Prize in 1962. Their discovery of the structure of DNA — the molecule that makes up every gene on earth, arranged in the shape of the now-famous double helix — allowed them, and later their colleagues, to deduce the mechanism by which DNA reproduced itself in generation after generation after generation. The process of DNA reproduction explained, in turn, how genes reproduce and how they maintain themselves endlessly, in precise and predictable order, along the chromosomes that characterize each living species. Watson and Crick's findings gave form to observations that Mendel could only begin to explain.

Suddenly geneticists were international heroes. In 1964, the year of Soviet Premier Nikita Khrushchev's fall from power, a private group of scientists convinced the Czechoslovakian government to move the Mendel statue again, this time to the more prominent of the monastery's two courtyards, where the original experiments had taken place. Instead of majestically gesturing toward the bustling Mendelplatz — now known by its Czech name, *Mendlovo náměstí* — with outstretched arms and an all-knowing gaze, the marble Mendel in this new location seemed humbler, milder, truer to the spirit that had guided the priest, the abbot, and the man. He seemed to both emerge from and retreat into the grove of evergreens in which the statue nestled. The grand marble staircase, gone for good, was replaced by an ordinary cement pedestal, keeping the statue just slightly higher than the people who came to see it.

After the Velvet Revolution of 1989 the new leaders of the democratic Czech Republic, as part of their clean sweep of old mistakes, offered to move the statue back to its original spot. *Mendlovo náměstí* was by then little more than a parched triangle of grass formed by the intersection of three tram tracks and flanked by the same four-story buildings that were there in 1910, which now housed beauty parlors, pubs, dress shops, and, in a few cases, dark deserted flats with cracked windows and an air of desolation.

Officials of the Mendelianum, a small museum of Mendelian artifacts located in the monastery's former dining hall, did not think relocation to such a barren site was a good idea. Mendel really belonged where he was, they said, in the courtyard, in a modest, green spot that still conveyed some dignity. The museum's scholars argued that if Mendel could have had a say in it, he would have chosen this place, out near the orangery, overlooking his garden.

So the statue now faces south, toward the plot of land where his pea plants grew. To disguise the loss of the ornate marble stairs, the monastery gardeners plant two rows of *Pisum sativum* along the statue's base every spring and again in the fall. On a morning in May the plants stand about six inches high, unsupported, their oval leaves gradually unfolding from each other like hands in easeful prayer. Their tendrils, with no stakes yet to grab hold of, reach over to the plants nearby, in what looks to be almost a strange botanical effort to create a kind of community — the very community that escaped Mendel, to his unending sadness and disappointment, throughout his lifetime.

Epilogue: Another Spring

> *A garden, then, is a finite place, in which a gardener*
> *... has created, working with or against nature, a*
> *plot whose intention is to provide pleasure. . . . And*
> *nature? Nature is what wins in the end.*
>
> — *The Gardener's Gripe Book*, Abby Adams

WHEN I FINALLY TRAVELED to pay homage to Mendel's monastery in the contemporary Czech city of Brno, my daughter accompanied me. Standing on the ramparts of the Špilberk Castle, the St. Thomas monastery far below us, we two were strangely happy. It had not been much fun to spend so much time in this grim and somber city, which still evokes, even in the brilliance of a May afternoon, a doleful Soviet chill. But for Mendel's true place in the history of genetics to make any sense, I knew I had to come here. There was no other way to arm myself with all the details it would take to turn Mendel inside out, back from a heroic founder into a flawed but brilliant human being.

So here I found myself, with my teenage daughter, on the trail of details. I walked the streets that Mendel walked, paced the garden in steady contemplation just as Mendel had. I fed my starved postmodern American senses: heard the dissonant clattering of the clock tower's bells atop the library wing; saw twilight vibrate with a color my daughter called "Tipton blue," the name of the stage lights that artificially re-create that rare sapphire shimmer; smelled the beery brewery and watched a European blackbird alight on a hundred-year-old tree.

Just days before we left for the Czech Republic, newspaper reports had appeared describing a new form of genetic engineering that seemed to subvert everything Mendel had worked for. In contrast to Mendel's attempts to learn from his plants' descendants just how inherited traits were passed on, contemporary plant geneticists were inventing ways to keep plants from passing on any traits at all. They were trying to make plants sterile, for no reason other than patent protection. The first step was the genetic modification of vegetable crops, inserting into them genes for disease resistance or increased productivity. But if these new strains of corn and soybeans were to reap a profit, scientists would have to find a way to prevent farmers from saving the previous year's seeds for cultivation the following spring, the way they did with normal crops. The solution was to insert into those bioengineered seeds one more gene: the gene for their own destruction. Once seeds were made sterile, farmers would have to come back to the biotech company every season to buy more.

The debate over the "terminator gene" in plants is the latest in a long string of debates over where modern genetics is leading us. Social questions have haunted this discipline for years, with Cassandras invoking chilly scenarios with every new bit of knowledge that geneticists have acquired. Today it seems as though some of their most dire warnings may be coming true. Consider the headlines. Dead men become fathers through their frozen sperm. Farm animals are genetically engineered to serve as factories for human pharmaceuticals, then are cloned so that more animals can carry on the same work in the identical way. Little girls are infused with modified viruses whose infectious qualities have been replaced with healthy genes the girls lacked at birth. The whole encyclopedia of the massive human genome is read, one bit of DNA after another, in perfect sequence, telling us with dreadful accuracy what it means to be normal.

This is where the early days of genetics, the days of Mendel and Nägeli, of Correns and De Vries, of Bateson and Weldon, have finally delivered us. It's enough to make you wonder what could possibly come next.

In the glare of the new twenty-first century, we know perhaps more than we need to — or even meant to learn — about our own genes and the genes of our fellow travelers on this planet. How much do we really want to know? I, at least, find that I want to know less about my own genetic legacy than I might have thought. My daughter and I, who traveled together back to a time when the units of inheritance were just starting to be revealed, each know personally the damage that can be wrought by a single deranged gene. Each of us may harbor a dominant gene that carries the progressive inherited disease that killed my father.

A test exists to tell us if we carry that gene. I have chosen not to take it. But my daughter would like to. If she finds out about her own genetic destiny, which she is undeniably entitled to do, I will learn something about my own destiny as well, even though I will not have chosen to. Because the disease gene is dominant, the only way she could get it is if I carry it, too. Should her test be positive, her knowledge would become my knowledge.

Knowledge is power, of course; that is what the last century of genetics has taught us. But knowledge can also be a catastrophe, as the history of humankind makes achingly clear. There is no such thing as a harmless putterer working quietly in his monastery garden. Mendel was asking what seemed on the surface to be innocent questions about how traits pass from parent to offspring, but his particular kind of genius led him to uncover the secrets of inheritance, the very mechanism of life itself. The more we discover — about how the cell does its work, how information is passed on from one generation to the next, and how that process can be intercepted, manipulated, and redirected toward some other, non-natural end — the more questions we must raise, one after another, in endless, recursive complexity. And after all the questions are answered, and new ones asked and answered once again, we may still be left wondering what exactly it was that we wanted to know.

Acknowledgments

Notes and Selected Readings

Index

ACKNOWLEDGMENTS

The most serious gardening I do would seem very strange to an onlooker, for it involves hours of walking round in circles, apparently doing nothing.

— *Garden Artistry,* Helen Dillon

What fun I've had investigating this story, and then telling it. For letting me have my fun, thanks go first to the Alfred P. Sloan Foundation, especially Doron Weber, who directs the program in the public understanding of science and technology. A Sloan grant allowed me to travel across the country and to Europe to interview Mendel scholars and walk in the footsteps of Gregor Mendel and his intellectual heirs.

Along the way I met some brilliant, open-hearted people: Will Provine of Cornell University, who first convinced me, with the simple spark of his enthusiasm, to redirect my thinking about the Mendel myth; Onno Meijer of the Free University of Amsterdam, who was my host during a frenetic, talk-stuffed day in Amsterdam that included two fabulous Mandarin meals; Lindley Darden of the University of Maryland, who let me audit her course on the history of modern biology in the spring of 1999; and Bob Olby of the University of Pittsburgh, with whom I spent a rain-soaked but fascinating afternoon.

In Europe, at the John Innes Centre in Norwich, England, where a large collection of Bateson's letters is housed in the John Innes Foundation Trustees Historical Collections, archivist Elizabeth Stratton was especially helpful, as were her supervisor, Ingrid Walton, and her assistant, Rachel Lewis, and her predecessor, Rosemary Harvey. At Cambridge University I had a fine lunch with Jim Secord, historian of science, and a fine coffee with Patrick Bateson, provost of King's College and, if I read the family tree correctly, William Bateson's first cousin twice removed. David Roe and Judy Goodman, amateur historians with the John Innes Society in the Merton Park section of southwest London, took me on a tour of Bateson's former residence (now the Rutlish School for Boys) and treated me to gossip about some of the area's most colorful residents, then and now. In Brno, Czech Republic, I spent nearly a week poring over old documents, haunting the St. Thomas monastery, and walking the streets of Brno with Anna Matalová, director of the Mendel Museum; and I spent an afternoon with Vítězslav Orel, the museum's director emeritus and one of the world's leading experts on Mendel's life and time.

Thanks to them all, and to Dennis Stevenson of the New York Botanical Gardens; Rob Cox of the American Philosophical Society library; Mike Ambrose of the John Innes Centre; Simon Mawer, author of *Mendel's Dwarf*; Naomi Davies and Godfrey Waller of the libraries at Cambridge; the librarians at the Smith College libraries and rare book room; Paul Bottino of the University of Maryland; Norm Weeden of Cornell University; Roger Blumberg, creator of the fabulous MendelWeb; Marek Havrda of Charles University in Prague; Jiri Sekerak, Marcela Sohajkova, Zdenek Polcak, and Helena Kostkova of the Mendel Museum; Chandak Sangoopta of the Wellcome Institute for the History of Medicine in London; and Jane Garmey, whose beautiful anthology, *The Writer in the Garden*, provided so many of the lovely quotes I used as chapter epigraphs. Thanks too to the Hamnett family for hosting me while I was in London, and to my old friends Anne Derbes and Bob Schwab for the liberal use of their beach house, currently my favorite writing spot on earth.

Writers who read all or part of the manuscript — Pat McNees, Erik Larson, Rob Kanigel, and others — proved once more, as if it needed proving, that they are good, good friends whom I've been lucky to know. Along with other members of our "mid-list group" of writers — especially Lynne Lamberg, Ed Regis, Aaron Levin, Jim Dilts, and Linda Lear — they gave just the right mix of advice and cheerleading. Other valued readers were Onno Meijer of Amsterdam, who ferried me incisive — and sometimes painful — comments via e-mail with his unique mix of bluntness, brilliance, and affection; Bob Donaldson, my daughters' favorite high school teacher and a true Renaissance man; and Clare Marantz, my mother, who helped more than she realized. My agent, Jean V. Naggar, offered much useful assistance, as did Peg Anderson, my copy editor, who had a light and lovely touch. And several friendly polyglots — Jason Owens and Laura Passin for German, Sharon Wolchik and Marketa Chromkova for Czech, Carolyn Rogers for Latin — provided speedy translations whenever I asked.

Anton Mueller, my editor at Houghton Mifflin, was simply amazing. He shepherded and shaped the book at every stage of its development, with an intelligence that was often, in every sense of the word, stunning. He has my unqualified thanks.

So does my little family. My older daughter, Jess, a history of science major at Smith College, was a true partner in this book. Among all the other dazzling things she is, Jess is a smart research assistant, editor, illustrator, and photographer, as well as a *simpatico* traveling companion. My younger daughter, Samantha, got me through many distracted periods by grounding me in the real world, the one that she so splendidly inhabits. And my husband, Jeff, was once again my most reliable sounding board. As he has always been, so he continues to be, through easy times and hard, in my writing and in every corner of my life: a steady source of comfort and my true, beloved friend.

NOTES AND SELECTED READINGS

> *Weeding is like housecleaning. . . . [It's] finicky work.*
> *It requires an overestimation of the importance of*
> *detail, a near-sighted view of things.*
>
> — *My Weeds: A Gardener's Botany,* Sara Stein

My bookshelves groan with books on Mendel and Mendelism, on genetics and evolution, on scientific discoveries and the nature of genius. My filing cabinets burst with photocopies of articles that relate to Mendel's paper, to the rediscovery, to the characters of De Vries and Correns and Bateson and Weldon, to what one paper calls "the reification of Mendel." A simple listing of all these books and articles, such as might appear in a scholarly text, would be misleading; there were only a few that I found myself returning to again and again. These are the sources that are listed under Selected Readings.

For a more comprehensive bibliography, look for this book's Web site at www.monkinthegarden.com. There you will find a listing of all those books stuffing my shelves and all the articles spilling out of my files. There you will also find the full text of some of the most important articles in this field, shown here with asterisks.

But before listing selected sources, I want to highlight the sources I relied on for each chapter and to embellish upon some of the thoughts I had room only to gloss over in the text.

Notes

Prologue: Spring 1900

The account of Bateson's train ride on May 8, 1900, is based on the recollections of his widow, Beatrice Bateson, in her introduction to his collected writings, *William Bateson, FRS, Naturalist*, p. 73. It has been contradicted by the historian Robert Olby in "William Bateson's Introduction of Mendelism to England: A Reassessment."

I believe Olby's reconstruction more than I believe Beatrice Bateson's. But I am fascinated by the legend built up around the supposed epiphany on the train, which has its own beautiful romance. I present the scene here as it was promulgated by the Batesons, and later in the book I revisit the scene from a different perspective.

1. In the Glasshouse

The biographical information, and the background information about Brünn and the St. Thomas monastery, are from the two official biographies of Mendel. The older one, from 1924, is by Hugo Iltis, *Life of Mendel*, which was originally published in German as *Gregor Johann Mendel, Leben, Werk und Wirkung*. This remained the standard biography of Mendel until the appearance in 1996 of a more modern one by Vítězslav Orel, *Gregor Mendel: The First Geneticist*. I used both of these books frequently, and I refer to them in the notes as Iltis and Orel.

Many of the details of monastery life in Mendel's day were provided by Anna Matalová, director of the Mendel Museum, during my visit there in May 1999.

2. Southern Exposure

Most of the details in this chapter are from Orel, with the students' recollections taken from Iltis. Mendel's waggish comment about the bishop is mentioned in the transcript of a performance by Richard M. Eakin of the University of California at Berkeley, dressed as Gregor Mendel in monk's garb from the 1860s. The performance was part of a symposium, "Science as a Way of Knowing — Genetics," presented at the annual meeting of the American Society of Zoologists, 27–30 December 1985, in Baltimore, Maryland. The transcript was published, as part of Eakin's "Great Scientists Speak Again" collection, in 1975 by the University of California Press; it was reprinted in *American Zoologist* 26 (1986): 749–52.

3. Between Science and God

Nietzsche announced that God is dead in *Thus Spake Zarathustra*, published in 1883. But God seems to have died at least three times in the history of Christendom, when interest in nature, inspired by religion, turned inadvertently against it. First, Aquinas stimulated the study of nature, thinking it would strengthen belief; the Galileo affair showed that this was not necessarily true. Second, in the Glorious Revolution natural science was hailed as the core of religion — but during the Enlightenment, God became unnecessary, since natural science was taken as a starting point quite separate from religious doctrine. Third, the theory that Darwin came up with in the late nineteenth century clashed directly with Church doctrine, as we see in Chapter 9.

The anthropocentric view of the earth and all its creatures is described in Peter J. Bowler, *Evolution: History of an Idea*, p. 53. I refer to Bowler's book often, especially in Chapter 9, simply as Bowler.

The suggestion that the discovery of entropy threatened this anthropocentric view was first made by my friend Roger Falcone, chairman of the department of physics at the University of California at Berkeley. Further details came from Clifford Truesdell, *The Tragicomical History of Thermodynamics, 1822–1854.*

4. Breakdown in Vienna

Information about Mendel's performance on his exam is from the chapter "Ploughed in an Examination" in Iltis. Mendel's complete zoology essay can be found at www.monkinthegarden.com.

My thanks to Onno Meijer for pointing out the "avalanche of numbers" (the term used by Ian Hacking in *The Emergence of Probability: A Philosophical Study of Early Ideas about Probability, Induction and Statistical Inference* [London and New York: Cambridge University Press, 1975]) that erupted during Mendel's time, including countings by Comte (knife killings), Galton (soldiers' heights), and Florence Nightingale (causes of death).

5. Back to the Garden

Many of the details in this chapter are from Orel, with a few provided, via e-mail or personal interviews, with Onno Meijer (about the progressive leanings of the Moravian Catholics in the late nineteenth century) and Anna Matalová (about Mendel's fondness for cucumbers and the way he took his weather readings). Details about the greenhouse and the conversion of the old glasshouse into an orangery are from an article by Orel, "The Building of Greenhouses in the Monastery Garden of Old Brno at the Time of Mendel's Experiments."

6. Crossings

The nonround peas are usually described as "wrinkled," but one of the most careful translators of Mendel's papers makes a convincing case that the word Mendel used, *kantig,* translates more accurately as "angular." "A search through English botanical texts confirms the accuracy of this description," writes the translator, Eva R. Sherwood: "the seeds are irregularly shaped, asymmetrically compressed, with smooth surfaces meeting at an angle." Because Robert Olby and other leading Mendel scholars consider Sherwood's translation to be the most authoritative, I have followed her lead in calling these peas angular.

Much of the information in this chapter comes from Iltis and two other books: Alain F. Corcos and Floyd V. Monaghan, *Gregor Mendel's Experiments on Plant Hybrids: A Guided Study,* and Robert Olby, *Origins of Mendelism.* Quotes from Mendel's letters to Nägeli are found in the invaluable collection coedited by Curt Stern and Eva R. Sherwood, *The Origin of Genetics: A Mendel Source Book.*

Onno Meijer took great pains to explain the long tradition of crossbreeding in an e-mail from Amsterdam in September 1999. "There is no doubt," he wrote, "that farmers [have] tried to improve their crops, by fertilizing or crossing, since time immemorial. One of the beliefs was that transmutation was possible — not only in witch-stories where you could be turned into a frog, but also in science. Note that alchemy was Newton's greatest occupation. . . . No reasonable mind from, say, the first settlers of Jericho up to the time of Mendel, could doubt that species could sometimes change over generations."

Regarding the accuracy of the English translations of Mendel's German-language paper, I talked to and read historians of science who have debated this issue, in particular the matter of *Merkmale* versus *Elemente.* I spent a lovely lunch discussing this, among many other things, with Simon Mawer, an English writer, biology teacher, and author of the scientific novel *Mendel's Dwarf,* during a trip he made to New York in July 1999. At his suggestion I counted the number of times *Merkmal* or *Merkmale* (singular and plural) and *Elemente* (always used in the plural) appeared in the original version of Mendel's paper, conveniently available for downloading — and computer searching — from the MendelWeb site (www.netspace.org/MendelWeb/).

Another word whose translation has been much discussed is *Entwicklung,* usually translated as "evolution," which can also mean "development." No one can be sure which one Mendel really meant: the development of an individual organism from conception to maturity, or the evolution of a population from one species into a new, closely related species.

7. First Harvest

The number of plants, peas, and blossoms (and therefore pods) that Mendel counted is calculated from his own estimate of examining "more than 10,000 plants," and from the estimates of others that Mendel pollinated an average of 4 flowers per plant, and from each plant got an average of 32.5 peas (this from Margaret Campbell, "Explanations of Mendel's Results," *Centaurus* 20 [1976]: 159–74).

Regarding Mendel's passion for numbers, Iltis described (p. 90) his odd arithmetical method for calling on boys in class. "Each boy had his own number, determined by his progress in the class. Mendel, fluttering the pages of a book, would select one of the numbers, 12, for instance. Then he would say, 'Twice 12 is 24, and 24 plus 12 is 36. I shall examine no. 36.'" Iltis believed Mendel's playfulness with numbers was a reflection of his efforts, during those days, to work out "the numerical ratios of inheritance." But it could just as easily have happened the other way. Maybe his search for ratios was the result, not the cause, of a love of counting and arithmetic that he had had all his life, long before he planted his first pea.

The probable timetable for Mendel's pea experiments is from Corcos and Monaghan's book (pp. 190–91) and from interviews with Robert Olby and Anna Matalová.

Once Mendel found the 3:1 ratio in his first few monohybrid crosses, he began to expect it in all subsequent F2 peas or plants. This might have led him to stop counting once the ratio was clearly established — or even to categorize certain peas or plants according to his expectations. There is, for instance, only the slightest difference between a pea that is yellowish green and one that is greenish yellow; into which pile should Mendel toss such peas, the green or the yellow? Because of his expectation of a 3:1 ratio, which became entrenched as a hypothesis as Mendel's *Pisum* work progressed, he may have made some decisions, either consciously or unconsciously, that made his data more convincing. These nearly perfect ratios were the source, in the 1930s, of suggestions that Mendel's data were too good to be true and that the priest — or a well-meaning assistant — must have fudged the data. For a thorough discussion of the controversy, the 1936 paper by statistician Ronald A. Fisher that started it all is available on line at www.monkinthegarden.com, as are two other relevant papers by Vítězslav Orel and Jan Sapp.

As in other chapters, biographical details in this chapter are mostly from Iltis and Orel, and quotes from Mendel's letters to Nägeli are from Stern and Sherwood, *Origin of Genetics*.

8. Eve's Homunculus

Maupertuis's theory of fluid semen is described in Bowler, p. 71, and also in Ernst Mayr, *The Growth of Biological Thought: Diversity, Evolution, and Intelligence,* pp. 328–29. Buffon's idiosyncrasies and beliefs are from Bowler, p. 73, and from William P. D. Wightman, *The Growth of Scientific Ideas* (New Haven: Yale University Press, 1951), p. 361. Blending inheritance is the subject of the chapter "Blending and Non-Blending Heredity: Darwin, Naudin, and Galton," in Olby, *Origins*, pp. 40–71.

As for real-world blending, Bob Donaldson, physics teacher at Blair High School in Silver Spring, Maryland, points out that in the classroom, mixing yellow and true blue paint will create black. The "blue" paint used in elementary schools is more accurately known as cyan.

9. The Flowering of Darwinism

Accounts of the showdown between Huxley and Wilberforce are from William Irvine, *Apes, Angels and Victorians: The Story of Darwin, Huxley, and Evolution* (New York: McGraw-Hill, 1955), and J. Vernon Jensen, *Thomas Henry Huxley: Communicating for Science* (Newark, Del.: University of Delaware Press, 1991). Details about Captain Fitzroy and Darwin's experiences on the *H.M.S. Beagle* are from Stephen J. Gould, *Ever Since Darwin: Reflections in Natural History* (New York: W. W. Norton, 1977), pp. 28–33. Most other details come from Bowler. Galton's antigemmule experiments on rabbits are described in Olby, *Origins*, p. 54; Steve Jones and Borin Van Loon, *Genetics for Beginners*, p. 11; and L. C. Dunn, *A Short History of Genetics* (Ames, Iowa: Iowa State University Press, 1965), p. 38.

10. Garden Reflections

One of Mendel's goals in beekeeping (described in Iltis, pp. 208–20) was to see whether his *Pisum* findings applied in animals. But bees proved an especially frustrating research model. Despite the elaborate fertilization cages Mendel designed and built, he was never able to limit which drones mated with the queen, and his pedigrees were therefore too inaccurate to allow him to draw any meaningful conclusions about bee hybridization.

Details about the London Exhibition, including building dimensions (pp. 33–34) and proposed amateur exhibits (pp. 51–53), are from John Timbs, *The Industry, Science, & Art of the Age: or, The International Exhibition of 1862* (London: Lockwood & Co., 1863).

The photo of the Moravian delegation at the Grand Hotel in Paris in July 1862 is from Orel, p. 196, and appears on line at www.monkinthegarden.com. Also at the Web site is a probable growing schedule for Mendel's trihybrid cross, between the summer of 1859 and the spring of 1862, taken from Corcos and Monaghan, *Gregor Mendel's Experiments,* pp. 192–95.

Mendel describes his findings on p. 22 of his paper (posted on the Mendel Web site). The numbers I use in the text are the collapsing of two smaller double-recessive backcrosses, one in which the double recessive acted as the male parent (providing the pollen), the other in which it was the female parent (providing the ovum). Mendel found that these so-called "reciprocal" crosses in every instance produced the same results no matter which type was the male and which the female. The crosses were like the commutative property of arithmetic, by which you can multiply two (or more) numbers in any order and always get the same result. On page 24 of his paper, Mendel states his hypothesis this way: "The pea hybrids form egg and pollen cells which, in their constitution, represent in equal numbers all constant forms which result from the combination of the characters united in fertilization."

11. Full Moon in February

The details of Mendel's lectures on February 8 and March 8, 1865, are from Iltis (pp. 80, 176, and 177) and Orel (pp. 89, 273, and 274).

Among the many apocryphal stories about Pythagoras (including the one about the dog) is that his reverence for beans was responsible for his death. When he was about sixty years old, the story goes, Pythagoras was being pursued by unnamed enemies in the Greek city of Metapontum. His pursuers chased him right up to the edge of a bean field, and there he stopped. Pythagoras refused to trample all those reincarnated babies and so was forced to face down his opponents — who murdered him.

In 1890 Oskar Hertwig gave the first full and completely correct description of meiosis. Hertwig's work, and the work of his predecessors Strasburger, Weismann, and Boveri, is described in Mayr, *Growth of Biological Thought*, pp. 761–64.

Several sources, including A. H. Sturtevant, *A History of Genetics* (New York: Harper & Row, 1965), p. 25, mention that many of the recovered reprints of Mendel's paper were found uncut, indicating that the recipients had not bothered to read them. But Anna Matalová insists that the original reprints were all cut before they were bound and distributed, meaning that the ones found in various libraries across the Continent could not have been uncut. I am relying on the more common story about the uncut reprints because it is a beautiful metaphor for the one thing we know for sure — that most of the scientists who received Mendel's reprint never bothered to read it.

The paths these reprints traveled are described in Theo J. Stomps, "On the Rediscovery of Mendel's Work by Hugo De Vries," *Journal of Heredity* 45 (1954): 294; Orel, p. 276; and Franz Weiling, "Fünf weitere Sonderdrucke der 'Versuche über Pflanzen-hybridin' J. G. Mendels Aufgetaucht," *Folia Mendeliana* 19 (1984): 257–63 (translated for me by Jason Owens).

12. *The Silence*

Information about the Mendel-Nägeli correspondence is from Iltis, pp. 167–69, 185, 190, 191, and 193. Quotes from Mendel's letters to Nägeli are from Stern and Sherwood, *Origin of Genetics*, pp. 62, 63, 77, 79, and 80.

Wilhelm Focke's description of Mendel's paper, and the fate of Focke's book as it made its way from Darwin's library to George Romanes's, are discussed in Augustine Brannigan, "The Reification of Mendel."

13. *"My Time Will Come"*

Mendel's letter to Nägeli about his eye ailment, which he attributes to "my own carelessness," appears in Stern and Sherwood, *Origin of Genetics*, pp. 86–87. Details about a typical Sunday playing skittles are from Iltis, p. 273. The items in Mendel's coat of arms and their meaning, the reason for his subsequent changes in his abbatial shield, and a description of the designs in the monastery library are from an interview with Anna Matalová in May 1999, when I visited the library and saw the ceiling decorations myself.

Mendel's careful description of the two-coned tornado includes a high level of detail, as recorded in Iltis, pp. 230 and 246. Iltis also records details of Mendel's decade-long fight over the monastery tax, in the chapter "The Struggle for the Right," pp. 253–72.

The comment about Mendel's paranoia, which extended even to his brethren, appeared in a letter from Anselm Rambousek — the man who would, within the year, succeed Mendel as abbot — to fellow monk Paul Křížkovský on May 8, 1883. Rambousek also commented that the abbot had "grown strikingly fatter," and he offered some unkind words about

Mendel's sister Theresia and her "stout" sixteen-year-old daughter, whom he described as a "walking regular tub." The letter is reprinted in an article by Orel, "Unknown Letters Relating to Mendel's State of Health," *Folia Mendeliana* 6 (1971): 268–69.

Mendel's personality is described variously in Iltis, pp. 219–20; in a chapter by Matalová, "Mendel's Personality — Still an Enigma?" in Orel and Matalová, *Gregor Mendel and the Foundation of Genetics;* and in an article by C. W. Eichling, "I Talked with Mendel," *Journal of Heredity* 33 (1942): 243–46. An analysis of his work with names as a form of mathematical linguistics, which at the time was a new branch of science, appears in Oldřich Ferdinand, "Mendel's Effort to Find Some Mathematical Laws in the Derivation of Names," *Folia Mendeliana* 1 (1966): 31–34.

Mendel's deathbed scene and obituary appear in Olby, pp. 105–6, and Niessl's eulogy is from Orel, p. 274. The Latin lines inscribed on the Augustinian memorial, which I visited in May 1999, are from Romans 14: 8, and were translated for me by Carolyn Rogers.

14. Synchronicity

The quotes from the myth of Demeter are from Geraldine McCaughrean, *Greek Myths,* illustrated by Emma Chichester Clark (New York: Margaret K. McElderry Books, 1993), p. 15. My neighbor, Jill Feasley, found this lovely book to read to her young children when she discovered, one evening in early June 1999, a wild evening primrose plant popping open in her front garden. She announced to the whole community, via a neighborhood e-mail listserve, that her yellow evening primroses were putting on a nightly display of synchronicity and that we were all invited to take part in the magic. For a few splendid weeks that June and July, dozens of us assembled every evening at about 8:45 to watch the fabulous unfurling taking place in Jill's little garden.

15. Mendel Redux

Much of the information about the rediscovery, including the "rediscovery papers" and letters from the three rediscoverers to the botanist H. F. Roberts, is from Roberts's classic book, *Plant Hybridization Before Mendel,* pp. 324–26.

Impressions of Correns's state of mind on April 21, 1900, come largely from Onno Meijer during an interview in October 1998. Meijer has made a study of the events of that day in 1900, including interviews with members of Correns's family.

A Dutch novel about homosexual love that came out in 1903, called

Pijpelijntjes (the title possibly a reference to the street language for oral sex), caused a huge antigay backlash throughout the Netherlands. The year *Pijpelijntjes* appeared was also the year Stomps went to work for De Vries, as described in Onno Meijer's article (which he summarized for me): "Hugo De Vries und Johann Gregor Mendel: Die Geschichte einer Verneining," *Folia Mendeliana* 21 (1986): 69–90. The book's appearance, and possibly the proximity of Stomps, caused De Vries such heartache that when he was offered a chance to move to the United States to work in the "fly room" of Thomas Hunt Morgan at Columbia University, he almost went. What kept him in Holland was the decision by the board of directors of the *Hortus Botanicus* to build De Vries his own research institute.

De Vries's ideas regarding mutation theory and intracellular pangenesis are summarized in Olby, *Origins,* pp. 11 and 112; Onno Meijer, "Hugo De Vries No Mendelian?"; and Corcos and Monaghan, "Was De Vries Really an Independent Discoverer of Mendel?" *Journal of Heredity* 76 (1985): 187–90.

Bateson's letter to Beatrice about De Vries is from Olby, *Origins,* p. 115; the text of his speech before the International Conference on Hybridization is from Beatrice Bateson, pp. 166 ff. Rolfle's mention of Mendel at that same conference is from Olby, *Origins,* p. 232.

The reason such a small percentage of Correns's maize hybrids exhibited a Mendelian ratio was not elucidated until two years later, when Correns discovered the phenomenon of selective fertilization. He found that an egg cell carrying the recessive sugary trait was less likely to be fertilized by pollen with the same trait than by pollen carrying the dominant starchy trait. This was later found to be related to the growth of the pollen tube, which can be affected by the genes it contains. This is described in Dunn, *A Short History of Genetics,* p. 102. Similarly, the story of the gene for colorlessness is from Sturtevant, *A History of Genetics,* p. 29.

Miescher's story is told in several places, but it is told best by Alfred E. Mirsky in "The Discovery of DNA," *Scientific American* 218, no. 6 (1968): 78–88. According to Mirsky, Miescher considered it essential to work with nuclein at low temperatures. His long hours, from five in the morning until late at night, working in unheated rooms through the fall and winter, have been blamed for his early death.

Erich von Tschermak is dismissed as a rediscoverer in Stern and Sherwood, *Origin of Genetics,* p. xi. His assessment of De Vries's motivation for dismissing Mendel is from his personal recollections, published as "The Rediscovery of Gregor Mendel's Work," *Journal of Heredity* 42 (1951): 163–71. De Vries's quote to Bateson regarding Mendel, made in a letter dated October 30, 1901, appears in William B. Provine, *The Origins of Theoretical Population Genetics,* p. 68.

16. The Monk's Bulldog

Details of William Bateson's life, including the text of many of his lectures and letters, are from Beatrice Bateson's memoir. Many of Bateson's letters are housed at the John Innes Centre archives in Norwich, England, and at the library of the American Philosophical Society in Philadelphia. R. C. Punnett's article, "Early Days of Genetics," *Heredity* 4 (1950): 7–8, provides a vivid description of the showdown at the British Association meeting in 1904 and the events leading up to it. Another book with lively descriptions of the Mendelian-biometrician controversy is Provine, *Origins of Theoretical Population Genetics.*

The story that Caroline Beatrice Durham wrote to win Bateson's heart was titled "At a Conversazione," *English Illustrated Magazine,* September 1895, pp. 551 ff.

17. A Death in Oxford

Most of the details of F. W. R. Weldon's life and death come from the lengthy obituary written about him by his friend and coeditor Karl Pearson in *Biometrika* 5, no. 1 (1906): 1–52.

18. Inventing Mendelism

The particulars about Bateson's personality, work, and writings are again from Beatrice Bateson's book and from Nicholas Mosley, *Hopeful Monsters* (London: Minerva, 1991); and David Lipset, *Gregory Bateson: Legacy of a Scientist.*

Information about terminology is from Elof Axel Carlson, *The Gene: A Critical History* (Philadelphia: W. B. Saunders, 1966). The history of the discovery of the chromosome and De Vries's mutation theory are described briefly in Mayr, *Growth of Biological Thought.*

Ideas about the transformation of Mendel into a larger-than-life "father of genetics" come from several sources, most helpfully Jan Sapp, "The Nine Lives of Gregor Mendel"; and Augustine Brannigan, "The Reification of Mendel." In addition, I learned much from personal interviews with Will Provine of Cornell University (August 1998), Onno Meijer of the Free University of Amsterdam (October 1998), and Jim Secord of Cambridge University (October 1998).

Stories about Thomas Morgan, including those of his stealing milk bottles and forgetting to ask after the health of his wife and new baby, and about the weird names for *Drosophila* mutants, are from Jonathan Weiner,

Time, Love, Memory: A Great Biologist and His Quest for the Origins of Behavior.

Until very recently, scientists estimated that the 3 billion or so letters in the human genome represented some 100,000 genes. This estimate was confirmed as the genomes of simple species were sequenced and found to have proportionately fewer genes: 19,000 in the worm knows as *C. elegans,* 13,000 in *Drosophila.* But in February 2001, when the entire human genome was sequenced and interpreted, it turned out that humans had not so very many more genes than these relatively primitive species. Today, the general estimate is that humans have only about 30,000 genes. "There is a lesson in humility in this," noted Eric Lander of MIT, one of the leading figures in the international publicly funded effort to sequence the human genome. "We only have twice as many genes as a fruit fly or lowly nematode worm. What a comedown." But Lander's associate, Francis Collins of the National Institutes of Health, thought this new finding made humans seem more magnificently complex than anyone had even imagined. "Because we have only about a third the number of genes that we expected, those genes must be particularly clever in carrying out their functions," he said on the television program *NewsHour* the night the new estimate was announced. "We're still just as complicated as before we figured this out, right? So it must mean that our genes have a certain elegant way of doing multiple tasks — more so, perhaps, than simpler organisms do."

19. A Statue in Mendelplatz

Bateson's confusion over the proper attire for the ceremony seems to have been a problem of translation. In a letter to Beatrice from Berlin on September 26, 1910, en route to Brünn, he wrote about the "rather vexing" question of his outfit. Apparently when he was told to wear *Gehrock,* he assumed it meant "my grey suit." But, as he realized when watching a play in Berlin, in which a character arrived in *Frock,* both *Frock* and *Gehrock* mean frock coat. Having brought the wrong suit, Bateson sent for his own frock coat — but it arrived at precisely ten o'clock on the morning of the unveiling, just as the Catholic mass was getting under way and too late for him to change into it. Following the suggestion of his hosts, he appeared at the ceremony wearing formal clothes instead.

Many of the quotes from Bateson about the statue and the ceremony are from A. G. Cock, "Bateson's Impressions at the Unveiling of the Mendel Monument at Brno in 1910."

Confusion among Brünners over just what Mendel "left" the town is noted in Iltis, p. 312. The fruit-fly-boiling order by Trofim Lysenko is from Robert F. Weaver and Philip W. Hedrick, *Genetics,* 3rd ed. (Dubuque, Iowa:

William C. Brown, 1997), p. 572. And the story of the Mendel statue being moved under cover of darkness in 1950 and 1964 comes from Anna Matalová, personal interview, May 1999.

Selected Readings

An asterisk indicates that the text can be found on the Internet at www.monkinthegarden.com.

Bateson, Beatrice. *William Bateson, FRS, Naturalist.* Cambridge, Eng.: Cambridge University Press, 1928.

Bowler, Peter J. *Evolution: The History of an Idea.* Rev. ed. Berkeley: University of California Press, 1989.

Brannigan, Augustine. "The Reification of Mendel." *Social Studies of Science* 9 (1979): 423–54.

Cock, A. G. "Bateson's Impressions at the Unveiling of the Mendel Monument at Brno in 1910." *Folia Mendeliana* 17 (1982): 217–23.

Corcos, Alain F., and Floyd V. Monaghan. *Gregor Mendel's Experiments on Plant Hybrids: A Guided Study.* Masterworks of Discovery Guided Studies of Great Texts in Science. New Brunswick, N.J.: Rutgers University Press, 1993.

Darden, Lindley. *Theory and Change in Science: Strategies from Mendelian Genetics.* New York: Oxford University Press, 1991.

Darwin, Charles. *On the Origin of Species: A Facsimile of the First Edition.* Cambridge, Mass.: Harvard University Press, 1964.

* DiTrocchio, F. "Mendel's Experiments: A Reinterpretation." *Journal of the History of Biology* 24 (1991): 485–519.

* Fisher, Ronald A. "Has Mendel's Work Been Rediscovered?" *Annals of Science* 1 (1936): 115–37.

Iltis, Hugo. *Life of Mendel.* Trans. Eden and Cedar Paul. London: George Allen & Unwin, 1932. Originally in German. *Gregor Johann Mendel, Leben, Werk und Wirkung.* Berlin: Julius Springer, 1924.

Isely, Duane. *One Hundred and One Botanists.* Ames, Iowa: Iowa State University Press, 1994.

Jones, Steve, and Borin Van Loon. *Genetics for Beginners.* Cambridge, Eng.: Icon Books, 1993.

Koestler, Arthur. *The Act of Creation.* New York: Dell, 1973.

Lipset, David. *Gregory Bateson: The Legacy of a Scientist.* Englewood Cliffs, N.J.: Prentice Hall, 1980.

Mawer, Simon. *Mendel's Dwarf.* New York: Harmony Books, 1998.

Mayr, Ernst. *The Growth of Biological Thought: Diversity, Evolution, and*

Intelligence. Cambridge, Mass.: Belknap Press of Harvard University Press, 1982.

* Meijer, Onno G. "Hugo De Vries No Mendelian?" *Annals of Science* 42 (1985): 189–232.

* Mendel, Gregor. *Experiments in Plant Hybridization.* Cambridge, Mass.: Harvard University Press, 1965. Originally published in German in *Verhandlungen des naturforschenden Vereines in Brünn,* Bd. IV für das Jahr 1865, Abhandlungen (1866): 3–47. First published in English in William Bateson, *Mendel's Principles of Heredity: A Defence.* Cambridge, Eng.: Cambridge University Press, 1902, pp. 40–95.

* Olby, Robert C. "Mendel No Mendelian?" *History of Science* 17 (1979): 53–72.

——. *Origins of Mendelism.* 2nd ed. Chicago: University of Chicago Press, 1985.

* ——. "William Bateson's Introduction of Mendelism to England: A Reassessment." *British Journal of History of Science* 20 (1987): 399–420.

Orel, Vítězslav. "The Building of Greenhouses in the Monastery Garden of Old Brno at the Time of Mendel's Experiments." *Folia Mendeliana* 10 (1975): 201–11.

——. *Gregor Mendel: The First Geneticist.* Oxford, Eng.: Oxford University Press, 1996.

* ——. "Will the Story of 'Too Good' Results of Mendel's Data Continue?" *BioScience* 18 (1968): 776–78.

Orel, Vítězslav, and Anna Matalová, eds. *Gregor Mendel and the Foundation of Genetics.* Brno, Czech Republic: Mendelianum of the Moravian Museum, 1983.

Provine, William. *The Origins of Theoretical Population Genetics.* Chicago: University of Chicago Press, 1971.

Roberts, H. F. *Plant Hybridization Before Mendel.* Princeton: Princeton University Press, 1929.

* Sapp, Jan. "The Nine Lives of Gregor Mendel." In H. E. Le Grand, ed., *Experimental Inquiries.* Netherlands: Kluwer Academic, 1990, pp. 137–66.

Stern, Curt, and Eva R. Sherwood. *The Origin of Genetics: A Mendel Source Book.* San Francisco: W. H. Freeman, 1966.

Taylor, Norman, ed. *Taylor's Encyclopedia of Gardening.* 4th ed. Boston: Houghton Mifflin, 1961.

Weiner, Jonathan. *Time, Love, Memory: A Great Biologist and His Quest for the Origins of Behavior.* New York: Alfred A. Knopf, 1999.

INDEX

Robin Marantz Henig is the author of six books, including *A Dancing Matrix: How Science Confronts Emerging Viruses,* for which she was named author of the year by the American Society of Journalists and Authors. She writes about science and medicine for such publications as the *New York Times Magazine, Civilization, Discover,* and *USA Today.*